MINI LABS ON GASES

*By Kathleen A. Kitzmann, Mercy High School, kakitzmann@mhsmi.org and
Kimberly Corliss, Waterford Mott High School, corliss1@pilot.msu.edu*

"Mini labs" are lab activities that can be done in a short period of time using a variety of activities that are set up around the lab room as stations. The number of stations depends on the room setup, the number of students, and the equipment available. Each of the authors has used mini labs to introduce the properties of gases to her chemistry classes. Both have found that the students eagerly participated in the activities and were able to build on the knowledge gained in these experiments.

There are many advantages to using mini labs:
(1)   Student interest is maintained since they are doing a variety of activities.
(2)   The individual activities usually take less than 5 minutes to complete.
(3)   Each individual activity usually emphasizes one key topic.
(4)   They can be used in any type of lab room and on any type of class schedule.
(5)   Not as much equipment is required. For example, one station might require a hot plate but only one is needed instead of 12 or 15.
(6)   If the mini labs are used to introduce a concept, the experiments may be referred to throughout the rest of the chapter.
(7)   Mini labs may be used to discover prior knowledge.
(8)   Opportunities are provided to develop the students' inquiry skills.
(9)   Once mini labs are set up one time, it is very easy to set them up again from year to year.
(10   Assessment of the students can range from simple to complex. Since these mini labs are often used as introductory activities", students are encouraged to make careful observations and to explain their ideas and reasoning.

Throughout the year, mini labs may be used with a variety of chemistry topics. Some possibilities include:
(1)   density activities;
(2)   physical and chemical changes;
(3)   types of reactions;
(4)   properties of solids and liquids;
(5)   properties of solutions; and
(6)   acids and bases.

In this paper, we are presenting some of the stations we used in our classes. At Mercy High School, there were nine stations set up at seven lab tables; at Waterford Mott High School, there were seven stations plus one additional demonstration. Three of our stations were the same; the others differed! For purposes of this article, we are presenting the three we had in common and one or two additional activities from each of us.

A sample set of student directions is provided first, followed by the materials needed for each station and some teacher notes. For additional information about any of these activities, please contact the authors. We hope these activities will inspire you to develop other Mini Labs for your chemistry classes.

GAS LAW MINI LABS
STUDENT DIRECTIONS

Station I: Cartesian Diver
Carefully squeeze the bottle on the table, and notice what happens to the dropper inside (a Cartesian diver). Watch the water level inside the dropper as the bottle is squeezed and released.

a. What happens to the pressure inside the bottle as you squeeze it?

b. Why does the diver rise and fall?

c. What do these observations tell you about the relationship between pressure and volume of a gas?

Station II: The Can Crusher
Add 3-5 mL of water to an empty soda can and place it on a hot plate. Turn the hot plate on high and observe the opening in the top of the can. As soon as you see steam coming out of the top of the can, use the beaker tongs provided and quickly invert the can in the tub of ice water.

a. What happens?

b. As the can is being heated, what replaces the air in the can?

c. Compare the gas pressure in the can with atmospheric pressure outside the can <u>before</u> and <u>after</u> you immerse it in the ice bath.

d. What causes the pressure inside the can to decrease and what does this indicate about the relationship between gas pressure and temperature?

Station III: Marshmallow Mafia (Mallows Under Pressure)
Remove the plunger on the syringe, and place a marshmallow inside. Cover the end of the syringe and press down on the plunger.

a. What happens?

Repeat with a new marshmallow, but this time place the marshmallow inside the syringe, push the plunger in until it is just touching the top, and then cover the end of the syringe. Pull the plunger out.

b. What happens this time?

c. Explain in detail what happens to the marshmallow as you change the position of the plunger, and what this tells you about the relationship between pressure and volume.

MSTA Journal • Spring 2003 • http://www.msta-mich.org

Station IV: Chemical Gas

Under the large clear container, there is one beaker (Beaker #1) containing a liquid. Please do not touch it. Take the empty beaker (Beaker #2) from the table, fill it about half way with water, and put a squirt of "solution A" into it. Place it under the large container next to Beaker #1. Observe for several minutes. After you are finished, remove Beaker #2, rinse it in the sink and leave it on the table for the next group. Leave Beaker #1 under the large container.

a.   What did you observe?

b.   What is the probable identity of "solution A"?

c.   What does this experiment tell you about the properties of gases?

Station V: Escaping Gases

Pick up and observe closely each balloon. Use all of your senses to observe them carefully.

a.   What did you observe?

b.   Why is there an odor coming from two of the balloons? What does this indicate about the molecules produc-ing the odor, and what does it indicate about the molecules making up the balloon's walls?

Station VI: Pop Your Top

Trial 1: Fill the film canister about half full of water. Drop in <u>half</u> of an antacid tablet. Quickly put the lid back on. STAND BACK! Time how long it takes the lid to pop off.

Trial 2: Change ONE variable, predict what you think will happen, and repeat the experiment.

a.   What did you observe in Trial 1?

b.   What did you change in Trial 2? What did you predict would happen? What actually happened?

Station VII: Volume Vs. Temperature

There are three balloons at this station, each at room temperature and inflated to the same size. Place one bal-loon in the beaker of ice water, the second in the beaker of boiling water, and the third on the table as a reference. Observe the effect of temperature on each of the balloons.

a.   What did you observe?

b.   What does this experiment show about the relationship between the volume of a gas and its temperature?

GAS LAW MINI LABS
TEACHER NOTES

Typically, we use each lab table or location in the room as a different activity. Students move freely from activity to activity, being sure that they complete each one. It is helpful to have a card labeling the station with a title or number. Directions can either be printed on the student worksheets, or on laminated cards at each table.

Station I: Cartesian Diver

The materials needed are several plastic 2-L pop bottles, filled almost to the top with water, and each containing a Cartesian Diver. Cartesian Divers may be made with plastic pipets and hex nuts. Flinn Scientific and other companies sell them. If a "Fizz Keeper" is available, it may be attached to one of the bottles and used to increase the pressure inside the bottle.

Many students will not realize the connection between density and the rise and fall of the diver. Students will say that the mass of WATER increased inside the diver, and the heavy water made it sink. It is helpful to remind them to consider changes on the GAS inside the diver. As the volume of air inside the diver is reduced, the density increases. It is helpful to have several bottle/diver set-ups ready to go, so all members of the group can make observations and play with the set up.

Answers to questions:
a. The pressure increases as you squeeze the bottle.
b. As the bottle is squeezed, the volume of air inside the diver decreases, the water level increases, and the diver sinks. As the sides of the bottle are released, the volume of air inside the diver increases and the diver rises.
c. Pressure and volume are inversely related.

Station II: Can Crusher

The materials needed are hot plates, a lot of pop cans, beaker tongs, large containers of ice water (beakers, dish pans, or similar containers). You will also want to provide a place for the crushed cans.

Some students will have to do this twice, because they put too much water in the can or do not wait long enough. Once the lab is running, you can ask students when they have finished with their can, to put another one on for the next group. This may be especially helpful on a shorter class schedule (fifty minute classes, for example) as it saves time and eliminates the "wait" for the water to heat up.

Answers to questions:
a. When the can is placed in the ice water, it collapses.
b. The air in the can is replaced with steam (water vapor).
c. After the can is filled with steam, the pressure inside the can equals the atmospheric pressure. When the can is placed in ice water, the pressure inside the can is less than atmospheric pressure, causing the can to "crush".
d. The pressure inside the can decreases because of the quick decrease in temperature. Temperature and gas pressure are directly related.

Station III: Marshmallow Mafia

The materials needed are plastic syringes (any size) with appropriately sized marshmallows. Miniature marshmallows fit 10 cc syringes; regular sized marshmallows work well in the larger syringes. Flinn Scientific sells large (1000 cc) syringes and caps. If both types are provided, the comparison is interesting. The end of the syringes may be covered with a finger, or with Luer Lock caps (small caps that fit tightly over the opening on the syringe). It is also fun for the students to draw faces on the marshmallows and watch the faces grow or shrivel inside the syringes.

Answers to questions:
a. The marshmallow shrivels.
b. The marshmallow expands.
c. Pressing down on the plunger decreases the volume of air in the syringe and increases the pressure. Since the marshmallow contains a lot of air, it shrinks as the pressure on it is increased. When the plunger is pulled out, the volume of air in the syringe increases which causes the pressure to decrease, and the marshmallow grows. Volume and pressure are inversely related.

Station IV: Chemical Gas

The materials needed are one large clear container (such as a 1000-mL beaker) and two smaller containers that will both fit under the large container (such as two 50-mL beakers). Fill one of the smaller beakers (labeled "Beaker 1") with ammonia. Label the second beaker "Beaker 2" and leave it empty on the table. "Solution A" is phenolphthalein and should be placed in a squirt bottle. This station will also need to have water available.

When the students place Beaker 2 (water + phenolphthalein) under the large container, next to Beaker 1, the indicator will turn pink. This demo can be repeated in class on the overhead projector. The heat of the overhead will cause the ammonia to vaporize slightly faster, and the time it takes for the reaction to occur will be less.

Note: Between classes, the teacher should lift up the large container to allow the ammonia vapors to escape. Otherwise, the change in indicator color will be almost instantaneous.

Answers to questions:
a. The solution in Beaker 2 gradually turns pink.
b. Solution A is most likely phenolphthalein.
c. Gases can diffuse through the air. Gases can react with other chemicals just like liquids and solids can. There was no "direct" contact between the solutions, so a gas must have come from Beaker 1 and entered Beaker 2 to cause the color change.

Station V: Escaping Gases

The materials needed are three balloons of different colors. Into two of the balloons squirt a milliliter or two of an extract with a distinct odor, like vanilla extract (in one), lemon extract (in the other), or any extract that you have available that might be recognizable to the students. One balloon has nothing in it. (Check these periodically; you may need to redo them during the day.)

Answers to questions:
a. (Answers will vary.)
b. The molecules producing the odor are gas molecules that are moving constantly and are tiny enough to pass through the walls of the balloon. The balloon has tiny holes in it that allow the molecules of gas to pass through.

Station VI: Pop Your Top

The materials needed are several film canisters (with lids), a source of water, stopwatches, and plenty of antacid tables. Each person or group performing the experiment will need at least one tablet. Practice with the film canisters ahead of time; some types work better than others. The first group or two that does this will "wake up" the rest of the class!

Answers to questions:

a. The top of the canister popped off after _?_ seconds.
b. Answers vary. Students should make a prediction and then give their actual results.

Station VII: Volume vs. Temperature

Three balloons are needed (different colors are preferable) and should be blown up to the same size, which should be less than the maximum allowed by the balloon and also small enough to fit into the beakers which will be used as the source of hot and cold water. A source of boiling water can be prepared by using a hot plate or burner and a large beaker of water. A beaker containing ice water should also be prepared. [If dry ice is available, this would be an even better source for the "cold" temperature balloon.]

Answers to questions:

a. The balloon placed in hot water expands; the balloon placed in ice water shrinks.
b. Volume and temperature are directly related.

Note: The authors presented a session entitled "Mini Labs in Chemistry" on Friday, March 14, 2003, at the MSTA Annual Conference in Grand Rapids, Michigan.

Flinn Scientific
ChemTopic™ Labs

Thermochemistry

Senior Editor

Irene Cesa
Flinn Scientific, Inc.
Batavia, IL

FLINN SCIENTIFIC INC.
"Your Safer Source for Science Supplies"
P.O. Box 219 • Batavia, IL 60510
1-800-452-1261 • www.flinnsci.com

ISBN 1-877991-78-3

Printed in the United States of America.

Table of Contents

Flinn ChemTopic™ Labs Series Preface
Lab Manuals Organized Around Key Content Areas in Chemistry

In conversations with chemistry teachers across the country we have heard a common concern. Teachers are frustrated with their current lab manuals, with experiments that are poorly designed and don't teach core concepts, with procedures that are rigid and inflexible and don't work. Teachers want greater flexibility in their choice of lab activities. As we further listened to experienced master teachers who regularly lead workshops and training seminars, another theme emerged. Master teachers mostly rely on collections of experiments and demonstrations they have put together themselves over the years. Some activities have been passed on like cherished family recipe cards from one teacher to another. Others have been adapted from one format to another to take advantage of new trends in microscale equipment and procedures, technology innovations, and discovery-based learning theory. In all cases the experiments and demonstrations have been fine-tuned based on real classroom experience.

Flinn Scientific has developed a series of lab manuals based on these "cherished recipe cards" of master teachers with proven excellence in both teaching students and training teachers. Created under the direction of an Advisory Board of award-winning chemistry teachers, each lab manual in the Flinn ChemTopic™ Labs series contains 4–6 student-tested experiments that focus on essential concepts and applications in a single content area. Each lab manual also contains 3–5 demonstrations that can be used to illustrate a chemical property, reaction, or relationship and will capture your students' attention. The experiments and demonstrations in the Flinn ChemTopic™ Labs series are enjoyable, highly focused, and will give students a real sense of accomplishment.

Laboratory experiments allow students to experience chemistry by doing chemistry. Experiments have been selected to provide students with a crystal-clear understanding of chemistry concepts and encourage students to think about these concepts critically and analytically. Well-written procedures are guaranteed to work. Reproducible data tables teach students how to organize their data so it is easily analyzed. Comprehensive teacher notes include a master materials list, solution preparation guide, complete sample data, and answers to all questions. Detailed lab hints and teaching tips show you how to conduct the experiment in your lab setting and how to identify student errors and misconceptions before students are led astray.

Chemical demonstrations provide another teaching tool for seeing chemistry in action. Because they are both visual and interactive, demonstrations allow teachers to take students on a journey of observation and understanding. Demonstrations provide additional resources to develop central themes and to magnify the power of observation in the classroom. Demonstrations using discrepant events challenge student misconceptions that must be broken down before new concepts can be learned. Use demonstrations to introduce new ideas, illustrate abstract concepts that cannot be covered in lab experiments, and provide a spark of excitement that will capture student interest and attention.

Safety, flexibility, and choice

Safety always comes first. Depend on Flinn Scientific to give you upfront advice and guidance on all safety and disposal issues. Each activity begins with a description of the hazards involved and the necessary safety precautions to avoid exposure to these hazards. Additional safety, handling, and disposal information is also contained in the teacher notes.

The selection of experiments and demonstrations in each Flinn ChemTopic™ Labs manual gives you the flexibility to choose activities that match the concepts your students need to learn. No single teacher will do all of the experiments and demonstrations with a single class. Some experiments and demonstrations may be more helpful with a beginning-level class, while others may be more suitable with an honors class. All of the experiments and demonstrations have been keyed to national content standards in science education.

Chemistry is an experimental science!

Whether they are practicing key measurement skills or searching for trends in the chemical properties of substances, all students will benefit from the opportunity to discover chemistry by doing chemistry. No matter what chemistry textbook you use in the classroom, Flinn ChemTopic™ Labs will help you give your students the necessary knowledge, skills, attitudes, and values to be successful in chemistry.

About the Curriculum Advisory Board

Flinn Scientific is honored to work with an outstanding group of dedicated chemistry teachers. The members of the Flinn ChemTopic Labs Advisory Board have generously contributed their proven experiments, demonstrations, and teaching tips to create these topic lab manuals. The wisdom, experience, creativity, and insight reflected in their lab activities guarantee that students who perform them will be more successful in learning chemistry. On behalf of all chemistry teachers, we thank the Advisory Board members for their service to the teaching profession and their dedication to the field of chemistry education.

Bob Becker teaches chemistry and AP chemistry at Kirkwood High School in Kirkwood, MO. Bob received his B.A. from Yale University and M.Ed. from Washington University and has 15 years of teaching experience. A well-known demonstrator, Bob has conducted more than 100 demonstration workshops across the U.S. and Canada and is currently a Team Leader for the Flinn Foundation Summer Workshop Program. His creative and unusual demonstrations have been published in the *Journal of Chemical Education,* the *Science Teacher,* and *Chem13 News.* Bob is the author of two books of chemical demonstrations, *Twenty Demonstrations Guaranteed to Knock Your Socks Off, Volumes I and II,* published by Flinn Scientific. Bob has been awarded the James Bryant Conant Award in High School Teaching from the American Chemical Society, the Regional Catalyst Award from the Chemical Manufacturers Association, and the Tandy Technology Scholar Award.

Kathleen J. Dombrink teaches chemistry and advanced-credit college chemistry at McCluer North High School in Florissant, MO. Kathleen received her B.A. in Chemistry from Holy Names College and M.S. in Chemistry from St. Louis University and has more than 30 years of teaching experience. Recognized for her strong support of professional development, Kathleen has been selected to participate in the Fulbright Memorial Fund Teacher Program in Japan and NEWMAST and Dow/NSTA Workshops. She served as co-editor of the inaugural issues of *Chem Matters* and was a Woodrow Wilson National Fellowship Foundation Chemistry Team Member for more than 10 years. Kathleen is currently a Team Leader for the Flinn Foundation Summer Workshop Program. Kathleen has been awarded the Midwest Regional Teaching Award from the American Chemical Society, the Tandy Technology Scholar Award, and a Regional Catalyst Award from the Chemical Manufacturers Association.

Robert Lewis teaches chemistry and AP chemistry at Downers Grove North High School in Downers Grove, IL. Robert received his B.A. from North Central College and M.A. from University of the South and has more than 25 years of teaching experience. He was a founding member of Weird Science, a group of chemistry teachers that has traveled throughout the country to stimulate teacher interest and enthusiasm for using demonstrations to teach science. Robert was a Chemistry Team Leader for the Woodrow Wilson National Fellowship Foundation and is currently a Team Leader for the Flinn Foundation Summer Workshop Program. Robert has received the Presidential Award, the James Bryant Conant Award in High School Teaching from the American Chemical Society, the Tandy Technology Scholar Award, a Regional Catalyst Award from the Chemical Manufacturers Association, and a Golden Apple Award from the State of Illinois.

John G. Little teaches chemistry and AP chemistry at St. Mary's High School in Stockton, CA. John received his B.S. and M.S. in Chemistry from University of the Pacific and has more than 35 years of teaching experience. Highly respected for his well-designed labs, John is the author of two lab manuals, *Chemistry Microscale Laboratory Manual* (DC Heath), and *Microscale Experiments for General Chemistry* (with Kenneth Williamson, Houghton Mifflin). He is also a contributing author to *Science Explorer* (Prentice Hall) and *World of Chemistry* (McDougal Littell). John served as a Chemistry Team Leader for the Woodrow Wilson National Fellowship Foundation from 1988 to 1997 and is currently a Team Leader for the Flinn Foundation Summer Workshop Program. He has been recognized for his dedicated teaching with the Tandy Technology Scholar Award and the Regional Catalyst Award from the Chemical Manufacturers Association.

Lee Marek teaches chemistry and AP chemistry at Naperville North High School in Naperville, IL. Lee received his B.S. in Chemical Engineering from the University of Illinois and M.S. degrees in both Physics and Chemistry from Roosevelt University. He has more than 30 years of teaching experience and is currently a Team Leader for the Flinn Foundation Summer Workshop Program. His students have won national recognition in the International Chemistry Olympiad, the Westinghouse Science Talent Search, and the Internet Science and Technology Fair. Lee was a founding member of ChemWest, a regional chemistry teachers alliance, and led this group for 14 years. Together with two other ChemWest members, Lee also founded Weird Science and has presented 500 demonstration and teaching workshops for more than 300,000 students and teachers across the country. Lee has performed science demonstrations on the *David Letterman Show* 20 times. Lee has received the Presidential Award, the James Bryant Conant Award in High School Teaching from the American Chemical Society, the National Catalyst Award from the Chemical Manufacturers Association, and the Tandy Technology Scholar Award.

John Mauch teaches chemistry and AP chemistry at Braintree High School in Braintree, MA. John received his B.A. in Chemistry from Whitworth College and M.A. in Curriculum and Education from Washington State University and has 25 years of teaching experience. John is an expert in "writing to learn" in the chemistry curriculum and in microscale chemistry. He is the author of two lab manuals, *Chemistry in Microscale, Volumes I and II* (Kendall/Hunt). He is also a dynamic and prolific demonstrator and workshop leader. John has presented the Flinn Scientific Chem Demo Extravaganza show at NSTA conventions for seven years and has conducted more than 100 workshops across the country. John was a Chemistry Team Member for the Woodrow Wilson National Fellowship Foundation program for four years and is currently a Team Leader for the Flinn Foundation Summer Workshop Program.

Dave Tanis is Associate Professor of Chemistry at Grand Valley State University in Allendale, MI. Dave received his B.S. in Physics and Mathematics from Calvin College and M.S. in Chemistry from Case Western Reserve University. He taught high school chemistry for 25 years before joining the staff at Grand Valley State University to direct a coalition for improving pre-college math and science education. Dave later joined the faculty at Grand Valley State University and currently teaches courses for pre-service teachers. The author of two laboratory manuals, Dave acknowledges the influence of early encounters with Hubert Alyea, Marge Gardner, Henry Heikkinen, and Bassam Shakhashiri in stimulating his long-standing interest in chemical demonstrations and experiments. Continuing this tradition of mentorship, Dave has led more than 40 one-week institutes for chemistry teachers and served as a Team Member for the Woodrow Wilson National Fellowship Foundation for 13 years. He is currently a Board Member for the Flinn Foundation Summer Workshop Program. Dave received the College Science Teacher of the Year Award from the Michigan Science Teachers Association.

Preface
Thermochemistry

There is an old saying that the only thing constant in life is change. This is certainly true in chemistry, where physical changes, chemical changes, and energy changes are all intertwined. The purpose of *Thermochemistry,* Volume 10 in the Flinn ChemTopic™ Labs series, is to provide high school chemistry teachers with laboratory activities that will lead students to a successful understanding of the role of heat and energy changes in chemistry.

How are processes characterized as exothermic or endothermic? What happens to the heat energy that is absorbed in an endothermic reaction? Can the amount of heat energy be measured? How much energy is released when food "burns" in the body? *Thermochemistry*—a collection of five experiments and five demonstrations—will help students answer these questions and appreciate the significant applications of thermochemistry in their daily lives.

Energy, heat, and temperature

Two demonstrations look at the relationship among these related but distinct concepts. "Colorful Heat" provides an effective visual aid to observe the difference between heat and temperature, while "Specific Heat" explores the definition of specific heat and how different substances can be used to store energy. Use the "Specific Heat" demonstration to introduce the factors involved in heat energy calculations.

Exothermic and endothermic reactions

Some reactions absorb heat as they proceed, while others release heat as they take place. In "Exploring Energy Changes," students observe the heat changes in physical and chemical reactions and then use technology to measure temperature changes over time in an endothermic or exothermic reaction. Two demonstrations—"The Cool Reaction" and "Flameless Ration Heaters"—also provide dramatic evidence of endothermic and exothermic reactions, respectively.

Calorimetry

Three experiments provide a range of choices for measuring heat transfer in physical and chemical processes. In "Measuring Energy Changes," students measure a heating curve for ice and water, and then use calorimetry to determine the amount of energy needed to melt ice. In "Discovering Instant Cold Packs," students design a calorimetry experiment to determine the enthalpy change that occurs when a "cold pack solid" dissolves in water. Finally, in "Measuring Calories," students use calorimetry to compare the amount of energy released when different snack foods burn.

Heats of reaction

"Heats of Reaction and Hess's Law" is an advanced microscale experiment that allows students to integrate their understanding of experiment and theory. Students determine the heats of reaction for reactions of magnesium metal and magnesium oxide and then apply Hess's law to calculate the heat of combustion of magnesium. Finally, the "Whoosh Bottle" provides a powerful demonstration of how much energy is released when fuels burn. Use this demonstration to introduce heat of reaction calculations and to illustrate the products formed in combustion reactions.

Incorporating technology

The use of technology for data collection and analysis is tailor-made for thermochemistry. "Exploring Energy Changes," the first experiment in this book, has been written as a technology-based lab activity. Specific instructions for adapting each activity to the use of calculator- or computer-based technology have also been included in the Supplementary Information of each experiment. In all cases, the use of technology should be viewed as an option, not a requirement.

Safety, flexibility and choice

The overlapping selection of experiments and demonstrations in *Thermochemistry* gives you the ability to cover the topics you feel are important in the safest, most effective manner possible. Beginning-level students will appreciate the opportunity to learn essential definitions not by memorizing them, but by seeing and feeling them, literally, in "Exploring Energy Changes." Students of all skill levels will benefit from the opportunity to discover calorimetry and connect it to their lives in the inquiry-based experiment, "Discovering Instant Cold Packs." Alternatively, the real-world application in "Measuring Calories" provides an effective learning exercise to make even the most resistant students sit up and take notice of what thermochemistry is all about. Finally, ambitious students will be challenged to put their knowledge of theory into practice in "Heats of Reaction and Hess's Law." Best of all, no matter what experiments and demonstrations you choose, your students are assured of success. Each experiment in *Thermochemistry* has been extensively tested and retested to make sure students will achieve meaningful results. Use the experiment summaries and concepts on the following pages to locate the concepts you want to teach and to choose experiments and demonstrations that will help you meet your goals.

Format and Features

Flinn ChemTopic™ Labs

All experiments and demonstrations in Flinn ChemTopic™ Labs are printed in a $10\frac{7}{8}''$ × 11″ format with a wide 2″ margin on the inside of each page. This reduces the printed area of each page to a standard $8\frac{1}{2}''$ × 11″ format suitable for copying.

The wide margin assures you the entire printed area can be easily reproduced without hurting the binding. The margin also provides a convenient place for teachers to add their own notes.

Concepts

Use these bulleted lists along with state and local standards, lesson plans, and your textbook to identify activities that will allow you to accomplish specific learning goals and objectives.

Background

A balanced source of information for students to understand why they are doing an experiment, what they are doing, and the types of questions the activity is designed to answer. This section is not meant to be exhaustive or to replace the students' textbook, but rather to identify the core concepts that should be covered before starting the lab.

Experiment Overview

Clearly defines the purpose of each experiment and how students will achieve this goal. Performing an experiment without a purpose is like getting travel directions without knowing your destination. It doesn't work, especially if you run into a roadblock and need to take a detour!

Pre-Lab Questions

Making sure that students are prepared for lab is the single most important element of lab safety. Pre-lab questions introduce new ideas or concepts, review key calculations, and reinforce safety recommendations. The pre-lab questions may be assigned as homework in preparation for lab or they may be used as the basis of a cooperative class activity before lab.

Materials

Lists chemical names, formulas, and amounts for all reagents—along with specific glassware and equipment—needed to perform the experiment as written. The material dispensing area is a main source of student delay, congestion, and accidents. Three dispensing stations per room are optimum for a class of 24 students working in pairs. To safely substitute different items for any of the recommended materials, refer to the *Lab Hints* section in each experiment or demonstration.

Safety Precautions

Instruct and warn students of the hazards associated with the materials or procedure and give specific recommendations and precautions to protect students from these hazards. Please review this section with students before beginning each experiment.

Procedure

This section contains a stepwise, easy-to-follow procedure, where each step generally refers to one action item. Contains reminders about safety and recording data where appropriate. For inquiry-based experiments the procedure may restate the experiment objective and give general guidelines for accomplishing this goal.

Data Tables

Data tables are included for each experiment and are referred to in the procedure. These are provided for convenience and to teach students the importance of keeping their data organized in order to analyze it. To encourage more student involvement, many teachers prefer to have students prepare their own data tables. This is an excellent pre-lab preparation activity—it ensures that students have read the procedure and are prepared for lab.

Post-Lab Questions or Data Analysis

This section takes students step-by-step through what they did, what they observed, and what it means. Meaningful questions encourage analysis and promote critical thinking skills. Where students need to perform calculations or graph data to analyze the results, these steps are also laid out sequentially and in order.

Format and Features

Teacher's Notes

Master Materials List

Lists the chemicals, glassware, and equipment needed to perform the experiment. All amounts have been calculated for a class of 30 students working in pairs. For smaller or larger class sizes or different working group sizes, please adjust the amounts proportionately.

Preparation of Solutions

Calculations and procedures are given for preparing all solutions, based on a class size of 30 students working in pairs. With the exception of particularly hazardous materials, the solution amounts generally include 10% extra to account for spillage and waste. Solution volumes may be rounded to convenient glassware sizes (100 mL, 250 mL, 500 mL, etc.)

Safety Precautions

Repeats the safety precautions given to the students and includes more detailed information relating to safety and handling of chemicals and glassware. Refers to Material Safety Data Sheets that should be available for all chemicals used in the laboratory.

Disposal

Refers to the current *Flinn Scientific Catalog/Reference Manual* for general guidelines and specific procedures governing the disposal of laboratory waste. Because we recommend that teachers review local regulations before beginning any disposal procedure, the information given in this section is for general reference purposes only. However, if a disposal step is included as part of the experimental procedure itself, then the specific solutions needed for disposal are described in this section.

Lab Hints

This section reveals common sources of student errors and misconceptions and where students are likely to need help. Identifies the recommended length of time needed to perform each experiment, suggests alternative chemicals and equipment that may be used, and reminds teachers about new techniques (filtration, pipeting, etc.) that should be reviewed prior to lab.

Teaching Tips

This section puts the experiment in perspective so that teachers can judge in more detail how and where a particular experiment will fit into their curriculum. Identifies the working assumptions about what students need to know in order to perform the experiment and answer the questions. Highlights historical background and applications-oriented information that may be of interest to students.

Sample Data

Complete, actual sample data obtained by performing the experiment exactly as written is included for each experiment. Student data will vary.

Answers to All Questions

Representative or typical answers to all questions. Includes sample calculations and graphs for all data analysis questions. Information of special interest to teachers only in this section is identified by the heading "Note to the teacher." Student answers will vary.

Look for these icons in the *Experiment Summaries and Concepts* section and in the *Teacher's Notes* of individual experiments to identify inquiry-, microscale-, and technology-based experiments, respectively.

Experiment Summaries and Concepts

Experiment

Exploring Energy Changes—Exothermic and Endothermic Reactions

Energy in the form of heat is exchanged in almost every chemical reaction or physical change in state. Some reactions absorb heat as they proceed, while others release heat as they take place. In this technology-based activity, students examine the heat changes in physical and chemical reactions and classify them as exothermic or endothermic. Students further investigate the amount of heat transfer in one of these reactions by measuring the resulting temperature change electronically as a function of time.

Measuring Energy Changes—Heat of Fusion

Our everyday experience tells us that energy in the form of heat is needed to melt ice or boil water. What happens to the heat energy that is absorbed in these endothermic processes? Can the amount of heat energy be determined? The purpose of this experiment is to study the temperature and heat changes that occur when ice melts. Students first measure a heating curve for ice, water, and steam, and then use the heat equation to determine the amount of heat needed to melt ice.

Discovering Instant Cold Packs—Heat of Solution

Instant cold packs are familiar first aid devices used to treat injuries. Most commercial cold packs consist of a plastic package containing a solid and an inner pouch filled with water. When the pack is activated, the solid dissolves in the water and produces a large temperature change. The purpose of this inquiry-based activity is to design a calorimetry experiment to determine the enthalpy change that occurs when a "cold pack solid" dissolves in water.

Measuring Calories—Energy Content of Food

All human activity requires "burning" food for energy. How much energy is released when food burns in the body? In this applied chemistry activity, students investigate the calorie content of different snack foods. Students use calorimetry to calculate the amount of heat energy released when different snack foods burn and identify patterns in the calorie content of foods.

Heats of Reaction and Hess's Law—Small-Scale Calorimetry

Burning magnesium in a Bunsen burner flame provides a dazzling demonstration of a combustion reaction. The intense flame produces blinding light and searing heat. The amount of heat produced in this reaction cannot be measured directly. In this microscale experiment, students use Hess's Law to determine the heat of reaction for the combustion of magnesium by an indirect method. The heats of reaction for magnesium and magnesium oxide reacting with hydrochloric acid are measured using special small-scale calorimeters and then combined algebraically with the known heat of formation of water to calculate the heat of combustion of magnesium.

Concepts

- Thermochemistry
- Heat
- Temperature
- Exothermic vs. endothermic
- System vs. surroundings

- Heat vs. temperature
- Exothermic vs. endothermic
- Heat of fusion
- Enthalpy change

- Enthalpy change
- Heat of solution
- Calorimetry
- Dependent and independent variables

- Combustion reaction
- Calorimetry
- Nutritional Calorie
- Calorie content of food

- Heat of reaction
- Heat of formation
- Hess's Law
- Calorimetry

Experiment Summaries and Concepts

Demonstration

Colorful Heat—Temperature versus Heat Demonstration

Students often use the terms heat and temperature interchangeably in their daily lives. Scientifically, however, these terms represent different quantities. The difference between heat and temperature is a key concept in thermochemistry. This demonstration provides a colorful illustration of the relationship between heat and temperature.

Specific Heat Chemical Demonstration

Three different metals of equal mass are heated to the temperature of boiling water. The metals are then added to three insulated foam cups, each containing the same amount of cold water. The resulting temperature changes are very different. This demonstration may be used on a qualitative level to illustrate the concept of specific heat, or it may be used to introduce quantitative heat and specific heat calculations.

The Cool Reaction—An Endothermic Demonstration

Ask students to think about chemical reactions, and they usually visualize exothermic reactions that produce heat, light, and often sound as well. What does an endothermic reaction look and feel like? In this classic demonstration, two solids are mixed, and the solid turns to a liquid—a freezing cold liquid. In less than two minutes, the temperature drops to –30 °C, cold enough to freeze water in contact with the reaction flask. "The Cool Reaction" is just that, cool and unusual!

Whoosh Bottle Chemical Demonstration

Combustion reactions represent the most important application of thermochemistry in our daily lives. The energy produced by the combustion of fossil fuels is used to heat our homes and power our vehicles. How much energy is released when fuels burn? In this classic "whoosh bottle" demonstration, students observe the dramatic "whoosh" of light, heat, and sound in the combustion of isopropyl alcohol and discover the products formed when organic compounds burn. A great way to introduce heat of combustion calculations!

Flameless Ration Heaters—An Applied Chemistry Demonstration

Developed by the United States Army for use by soldiers in the field, flameless ration heaters (FRH's) contain a composite magnesium/iron alloy material that reacts exothermically upon activation with water. The amount of heat released is sufficient to increase the temperature of food from room temperature to about 75 °C. In this interesting real-world demonstration of the applications of thermochemistry, students observe the heat of reaction for activation of an FRH with water and deduce the composition of the metallic heating pad based on qualitative chemical tests.

Concepts

- Temperature
- Heat

- Heat
- Specific heat
- Calorimetry

- Endothermic reaction
- Heat of reaction
- Enthalpy
- Entropy

- Exothermic reaction
- Combustion reaction
- Heat of combustion

- Exothermic reaction
- Heat of reaction

Teacher Notes

Exploring Energy Changes
Exothermic and Endothermic Reactions

Introduction

The story of chemistry is the story of change—physical change, chemical change, and energy change. Energy in the form of heat is exchanged in almost every chemical reaction or change in state. Some reactions require heat in order to proceed. Other reactions release heat as they take place. In this experiment, we will investigate several processes in order to get a feel for the amount of heat absorbed or released in physical and chemical reactions.

Concepts

- Thermochemistry
- Heat
- Exothermic vs. endothermic

- Energy
- Temperature
- System vs. surroundings

Background

Thermochemistry is the study of heat changes that take place in a change of state or a chemical reaction—heat energy is either absorbed or released. If a process releases energy in the form of heat, the process is called exothermic. A process that absorbs heat is called endothermic. How do we observe or measure the heat change that occurs in a physical or chemical change?

Heat is defined as the energy transferred from one object to another due to a difference in temperature. We do not observe or measure heat directly—we measure the temperature change that accompanies heat transfer. In a chemical reaction it is often not possible to measure the temperature of the reactants or products themselves. Instead, we measure the temperature change of their surroundings.

The difference between the system and the surroundings is a key concept in thermochemistry. The *system* consists of the reactants and products of the reaction. The solvent, the container, the atmosphere above the reaction (in other words, the rest of the universe) are considered the *surroundings*. Heat may be transferred from the system to the surroundings (the temperature of the surroundings will increase) or from the surroundings to the system (the temperature of the surroundings will decrease).

When a system releases heat to the surroundings during a reaction, the temperature of the surroundings increases and the reaction container feels warm to the touch. This is an *exothermic reaction*—the prefix *exo-* means "out of" and the root *thermic* means heat. Heat flows out of the sytem. An example of an exothermic reaction is the combustion of propane (C_3H_8) in a barbecue grill to produce carbon dioxide, water, and heat. *Equation 1* gives the chemical equation for this reaction; notice that heat appears on the product side in the equation for an exothermic reaction.

$$C_3H_8(g) + 5O_2(g) \rightarrow 3CO_2(g) + 4H_2O(g) + \text{heat} \qquad \textit{Equation 1}$$

The difference between heat and temperature is not obvious to all students. Try the "Colorful Heat" demonstration in this lab manual to illustrate the relationship between heat and temperature.

When a system absorbs heat from the surroundings during a reaction, the temperature of the surroundings decreases and the reaction container feels cold to the touch. This is an *endothermic reaction,* where the prefix *endo-* means "into." Heat flows into the system. A familiar example of an endothermic process is the melting of ice. Solid water (ice) needs heat energy to break the forces holding the molecules together in the solid state. This physical change is represented by *Equation 2;* notice that heat appears on the reactant side in the equation for an endothermic reaction.

$$H_2O(s) \; + \; heat \; \rightarrow \; H_2O(l) \hspace{4cm} Equation \; 2$$

Experiment Overview

The purpose of this experiment is to examine the heat changes in physical and chemical processes and to classify them as exothermic or endothermic. In Part A, three reactions are carried out in heavy-duty, zipper-lock plastic bags. The bags make it easy to observe and feel the heat changes that occur. In Part B, the extent of heat transfer in one of these reactions will be investigated by measuring the resulting temperature change. The reaction will be carried out in an insulated foam cup and the temperature of the solution will be measured as a function of time.

Pre-Lab Questions

1. Read the entire procedure and the recommended safety precautions. What hazards are associated with the use of hydrochloric acid in the lab? How can these hazards be reduced?

2. Classify each of the following processes as a physical change or a chemical change and as an exothermic or endothermic reaction.

    (a) Sugar is dissolved in water in a test tube and the test tube feels cold.

    (b) Gasoline is burned in a car engine.

    (c) Water is converted to steam according to the equation $H_2O(l) \; + \; heat \; \rightarrow \; H_2O(g)$.

3. Two solutions, hydrochloric acid and sodium hydroxide, were mixed and the temperature of the resulting solution was measured as a function of time. The following graph was recorded. Is the reaction between hydrochloric acid and sodium hydroxide exothermic or endothermic?

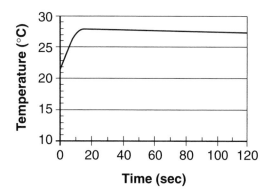

Teacher Notes

Materials

Ammonium chloride, NH_4Cl, 8–10 g	Balance, centigram (0.01 g precision)
Calcium chloride, $CaCl_2$, 12–14 g	Beaker, 400-mL
Hydrochloric acid solution, HCl, 1 M, 40 mL	Graduated cylinders, 10- and 50-mL
Sodium bicarbonate, $NaHCO_3$, 3–5 g	Pen for labeling
Water, distilled or deionized	Sealable, zipper-lock plastic bag, 1
Temperature sensor or thermometer	Spatula
Computer interface system (LabPro™)*	Insulated foam (Styrofoam®) cup, 6 oz, 1
Computer or calculator for data collection*	Weighing dishes or small beakers, 4
Data collection software (LoggerPro™)*	
*Optional	

Safety Precautions

Hydrochloric acid solution is toxic by ingestion or inhalation and is corrosive to eyes and skin. Avoid contact with eyes and skin. Notify the teacher and clean up all spills immediately with large amounts of water. Ammonium chloride and calcium chloride are slightly toxic by ingestion. Wear chemical splash goggles and chemical-resistant gloves and apron. Wash hands thoroughly with soap and water before leaving the laboratory.

Procedure

Part A. Observing Heat Changes

1. Obtain 3 weighing dishes or small beakers and label them A–C.

2. Weigh out the appropriate amount of solid into each weighing dish, according to the following table. Record the exact mass of each solid in Data Table A.

Weighing Dish	A	B	C
Solid	Ammonium chloride	Calcium chloride	Sodium bicarbonate
Mass	2–3 g	3–4 g	1–2 g

3. Open a zipper-lock plastic bag and pour the solid from A into the bottom of the bag. Tilt the bag so all the solid falls into one corner of the bag, then lay the bag flat on the table.

4. Measure 10 mL of distilled water in a graduated cylinder and pour the water into the bag, trying to pool the water in the upper third of the bag.

5. Close the zipper-lock bag and gently squeeze the bag to mix the solid and liquid contents.

6. Feel the side of the bag where the reaction is occurring and record whether the bag feels hot or cold to the touch. Observe what happens as the contents mix. Record all observations in Data Table A.

7. Wash the contents of the bag down the drain with excess water. Rinse the inside of the bag with distilled water and dry it using a paper towel.

8. Repeat steps 3–6 using sample B and 10 mL of distilled water.

9. Wash the contents of the bag down the drain with excess water. Rinse the inside of the bag with distilled water and dry it using a paper towel.

Many different solids may be used to demonstrate endothermic and exothermic dissolving reactions. Ammonium nitrate may be used as a substitute for ammonium chloride in this experiment, while sodium carbonate may be used as a substitute for calcium chloride.

10. Repeat steps 3–6 using sample C and 10 mL of 1 M hydrochloric acid solution.

11. Wash the contents of the bag down the drain with excess water. Dispose of the bag as instructed by your teacher.

Part B. Measuring Temperature vs. Time

Your teacher will assign you and your group one of the reactions from Part A to study in more detail. Record the identity of the reaction assigned to you in Data Table B. Use the following table to determine the required amount of solid and liquid for steps 13 and 14.

Reaction	Solid (g)	Liquid (mL)
A	Ammonium chloride (6–7 g)	Distilled water (30 mL)
B	Calcium chloride (9–10 g)	Distilled water (30 mL)
C	Sodium bicarbonate (2–3 g)	Hydrochloric acid, 1 M (30 mL)

12. Set an empty, dry Styrofoam cup into a 400-mL beaker so that the cup is stable and will not tip over.

13. Weigh out the required amount of solid in a weighing dish and record the identity and exact mass of the solid in Data Table B.

14. Measure 30.0 mL of the appropriate liquid in a graduated cylinder and pour the liquid into the Styrofoam cup. Record the identity and volume of the liquid in Data Table B.

15. Plug a temperature sensor into the interface system.

16. Open and format a graph in the data collection software so that the y-axis reads temperature in degrees Celsius. Set the minimum and maximum temperature values at 0 and 40 °C, respectively.

17. The x-axis should be set for time in seconds. Set the minimum and maximum time values at 0 and 240 sec, respectively.

18. Set the time interval to take a temperature reading every 10 seconds.

19. Place the temperature probe in the liquid in the Styrofoam cup.

20. Wait one minute (to allow the temperature sensor to become acclimated to the liquid temperature), then press start to begin collecting data. *Immediately add the solid from the weighing dish into the Styrofoam cup and gently mix the contents using a stirring rod.*

21. The system will automatically record data for the allotted time (240 sec), then stop.

22. Remove the sensor from the Styrofoam cup and rinse it with distilled or deionized water. Wash the contents of the cup down the drain with excess water.

23. If possible, obtain a printout of the data table and graph from the computer.

24. *Complete Data Table B:* Using the data from the computer table and graph of temperature vs. time, record the initial temperature of the solution (before adding solid) and the maximum or minimum temperature obtained after mixing.

It may be easier to supply each group with three plastic bags for Part A. Ordinary household-type bags will work fine and will save time if students do not have to wash and dry the bags between reactions in Part A.

Teacher Notes

Name: _____

Class/Lab Period: _____

Exploring Energy Changes

Data Table A. *Observing Heat Changes*

Reaction	Solid + Liquid	Mass of Solid (g)	Observations
A	$NH_4Cl(s) + H_2O(l)$		
B	$CaCl_2(s) + H_2O(l)$		
C	$NaHCO_3(s) + HCl(aq)$		

Data Table B. *Measuring Temperature vs. Time*

Assigned Reaction	
Identity of Solid	
Mass of Solid (g)	
Identity of Liquid	
Volume of Liquid (mL)	
Initial Temperature (°C)	
Maximum or Minimum Temperature (°C)	

Post-Lab Questions

Attach the printout of the data table and graph for Part B to your lab report.

1. Complete the following Results Table to indicate whether each reaction in Part A represents a physical or chemical change and whether it is exothermic or endothermic.

Reaction	Physical or Chemical Change?	Exothermic or Endothermic?
A		
B		
C		

2. A chemical change involves a change in the composition of matter—the formation of a new chemical substance (product) with physical and chemical properties different from those of the reactants. Describe the evidence used to decide if any of the processes in Part A were chemical changes.

3. Did you observe any qualitative differences in the amount of heat generated in the reactions that were characterized as endothermic in Part A?

4. Consider Reaction A: Was energy released or absorbed by the reactants in this system? When you *touched* the reaction container (the plastic bag) was energy being released or absorbed by your *hand?*

5. Write a balanced equation for each of the processes in Part A. Remember to include heat on the reactant or product side, as appropriate.

6. In Part B, was the temperature that was measured a part of the system or the surroundings? How long did it take for the maximum or minimum temperature to be reached?

7. Describe in words the temperature versus time graph that was recorded in Part B. Be as specific as possible.

8. Complete the following sentence to summarize the observations and conclusions for the reaction in Part B: The reaction of _____ with _____ is an (exothermic/endothermic) process in which energy was (absorbed/released) by the system and the temperature of the surroundings (increased/decreased).

Teacher's Notes
Exploring Energy Changes

Master Materials List *(for a class of 30 students working in pairs)*

Ammonum chloride, NH_4Cl, 60–80 g

Calcium chloride, $CaCl_2$, 90–110 g

Hydrochloric acid solution, HCl, 1 M, 300 mL

Sodium bicarbonate, $NaHCO_3$, 25–40 g

Water, distilled or deionized

Temperature sensors or thermometers

Computer interface systems (LabPro), 15*

Computers or calculator for data collection*

Data collection software (LoggerPro)*

*Optional. See the *Lab Hints* section for other options.

Balance, centigram (0.01 g precision), 3

Beakers, 400-mL, 15

Graduated cylinders, 10- and 50-mL, 15 each

Pens for labeling

Sealable, zipper-lock plastic bags, 15

Spatulas, 15

Insulated foam (Styrofoam®) cups, 6 oz, 15

Weighing dishes or small beakers, 60

Preparation of Solutions *(for a class of 30 students working in pairs)*

Hydrochloric acid, 1 M: Add about 250 mL of distilled or deionized water to a 500-mL flask. Using a funnel, carefully add 41 mL of 12 M (concentrated) HCl to the flask. Stir to mix, then dilute to 500 mL with distilled water. *Remember:* Always add acid to water.

Safety Precautions

Hydrochloric acid solution is toxic by ingestion or inhalation and is corrosive to eyes and skin. Ammonium chloride and calcium chloride are slightly toxic by ingestion. Avoid contact of all chemicals with eyes and skin. Wear chemical splash goggles and chemical-resistant gloves and apron. Wash hands thoroughly with soap and water before leaving the laboratory. Clean up all spills immediately with large amounts of water. Please consult current Material Safety Data Sheets for additional safety, handling, and disposal information.

Disposal

Consult your current *Flinn Scientific Catalog/Reference Manual* for general guidelines and specific procedures governing the disposal of laboratory waste. All solutions from Parts A and B may be rinsed down the drain with excess water according to Flinn Suggested Disposal Method #26b.

Lab Hints

- This experiment is designed as an exploratory, qualitative introduction to thermochemistry. The emphasis is not on calculations, but rather on getting a feel, literally, for the concept of heat transfer in physical and chemical processes. The experiment can reasonably be completed in one 50-minute lab period.

The use of technology is an option, not a requirement. See the Lab Hints section for how to do this experiment without using technology. For more information on data collection systems such as LabPro and CBL-2, please see your current Flinn Scientific Catalog/ Reference Manual.

- Heavy-duty, sealable plastic bags provide convenient reaction containers for Part A. It is much easier for students to feel the heat changes in these plastic bags than in conventional lab glassware such as beakers, Erlenmeyer flasks, or graduated cylinders. The reactions are very quick and the bags are reusable. Normal grocery store zipper-lock plastic bags will work fine. Be sure to try these household-type bags first before using them in this experiment. To save time, give each student group three bags—that way students do not have to wash, rinse, and dry the bags between trials.

- The use of computer- or calculator-based technology for data collection and analysis is tailor-made for thermochemistry experiments. Specific instructions have been given in this first experiment in the thermochemistry series for electronic measurements of temperature versus time. Introducing the use of technology in this exploratory lab, rather than in a more advanced experiment such as Hess's Law, means students become accustomed to the technology earlier and in a comfortable environment.

- The use of technology is not essential to the success of this experiment. Temperature versus time measurements in Part B may also be made using thermometers. Digital thermometers are preferred over glass-bulb thermometers because they provide direct readings, update every second, and give ±0.1 °C precision. While glass-bulb thermometers may also be used, they suffer from the fact that the 1 °C divisions make it difficult for many students to quickly approximate to the nearest 0.5 °C. Temperature readings should be obtained every 15–20 seconds if glass-bulb thermometers are used.

- The temperature decrease observed upon reaction of sodium bicarbonate with hydrochloric acid is a classic unexpected event. Given the spontaneous fizzing, many students would predict that the temperature should increase not decrease.

- At the conclusion of Part B, have students post their results on the board. Compare variations in the maximum or minimum recorded temperature for Reactions A, B, and C and discuss the source of these variations and how these variations might be controlled in a quantitative experiment.

Teaching Tips

- There are two challenging conceptual hurdles students must overcome in thermochemistry—the difference between *heat* and *temperature* and the definition of the *system* versus the *surroundings*. For a reaction taking place in solution, students must realize that the liquid, the solvent, is not directly involved in the reaction. It is part of the medium, the surroundings. Reactions are generally classified as exothermic or endothermic based on the temperature change in the surroundings, which is opposite in sign to that of the system. Thus, if the temperature of the surroundings increases, it is because the energy of the system has decreased. The temperature of the system itself is often inaccessible.

- The "Colorful Heat" demonstration in this Flinn ChemTopic™ Labs manual illustrates the difference between heat and temperature. A constant amount of food coloring (heat) is added to different volume amounts of water. The resulting color intensity (temperature) depends not only on the amount of heat added but also on the amount of material absorbing the heat.

The technology instructions in Part B are general enough that they may be used for other interface systems, such as the CBL-2 and graphing calculators.

Teacher Notes

- Sufficient data is collected in Part B to allow students to calculate the amount of energy transferred to or from the surroundings. The amount of heat (q) transferred from the system to the surroundings depends on three factors: the mass (m) of the surroundings, its specific heat (s), and the temperature change (ΔT), according to the following equation. The units are shown in parentheses.

$$q \text{ (J)} = m \text{ (g)} \times s \text{ (J/g·°C)} \times \Delta T \text{ (°C)}$$

- The heat of solution for ammonium chloride dissolving in water in Reaction A was calculated using the sample data shown in Data Table B and was found to be 335 J/g. (This compares with the literature value of 277 J/g.) *Note:* The specific heat (s) and density of the final solution were assumed to be the same as those of water (s = 4.18 J/g·°C and density = 1.0 g/mL). The mass of the surroundings was assumed to be the sum of the initial masses of water AND ammonium chloride. The literature value for the heat of solution of calcium chloride is –747 J/g.

Answers to Pre-Lab Questions *(Student answers will vary.)*

1. Read the entire procedure and the recommended safety precautions. What hazards are associated with the use of hydrochloric acid in the lab? How can these hazards be avoided?

   Hydrochloric acid is a corrosive liquid and can damage and burn skin and eyes. Avoid contact with skin and eyes. Wear chemical spash goggles and chemical-resistant gloves and apron. Notify the teacher and clean up all spills immediately.

2. Classify each of the following processes as a physical change or a chemical change and as an exothermic or endothermic reaction.

   (a) Sugar is dissolved in water in a test tube and the test tube feels cold.

   Physical change—the composition of sugar is unchanged after dissolving in water. Endothermic—absorbs heat energy from the surroundings so that the test tube feels cold.

   (b) Gasoline is burned in a car engine.

   Chemical change—gasoline is broken down to carbon dioxide and water. Exothermic—heat energy is produced and the engine gets very hot. **Note:** *Much of the chemical energy of gasoline, of course, is also converted to mechanical energy in the pistons.*

   (c) Water is converted to steam according to the equation $H_2O(l) + heat \rightarrow H_2O(g)$.

   Physical change of state—evaporation or boiling. Endothermic—heat appears on the reactant side in the equation.

Answers to Pre-Lab Questions continue on page 10.

3. Two solutions, hydrochloric acid and sodium hydroxide, were mixed and the temperature of the resulting solution was measured as a function of time. The following graph was recorded. Is the reaction between hydrochloric acid and sodium hydroxide exothermic or endothermic?

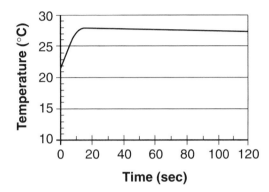

Exothermic reaction—the temperature of the solution (the surroundings!) increased from 21 °C to about 28 °C. This temperature increase is due to the heat energy released by the system—the reactants (HCl and NaOH) and products (NaCl and H_2O)—and absorbed by the water in solution.

Sample Data

Student data will vary.

Data Table A. *Observing Heat Changes*

Reaction	Solid + Liquid	Mass of Solid (g)	Observations
A	$NH_4Cl(s) + H_2O(l)$	2.52	The white solid slowly dissolved in the liquid. The bag felt very cold to the touch. Not all of the solid dissolved. The final mixture was a chalky white suspension.
B	$CaCl_2(s) + H_2O(l)$	3.05	The white solid dissolved almost instantly. The bag felt hot to the touch. All of the solid dissolved and the final mixture was a colorless solution.
C	$NaHCO_3(s) + HCl(aq)$	1.36	Vigorous frothing; white solid disappeared instantly and mixture fizzed. Lots of bubbles produced, then subsided. The bag felt slightly cold to the touch. No solid remained—final mixture was a colorless, bubbly solution.

Data Table B. *Measuring Temperature vs. Time*

Assigned Reaction	A	B	C
Identity of solid	NH_4Cl	$CaCl_2$	$NaHCO_3$
Mass of solid (g)	6.24	9.21	2.29
Identity of liquid	H_2O	H_2O	HCl
Volume of liquid (mL)	30	30	30
Initial temperature (°C)	18.9	19.3	19.7
Maximum or minimum temperature (°C)	5.1	38.7	14.1

Teacher's Notes

Sample Graphs. *Measuring Temperature versus Time*

Ammonium Chloride/Water

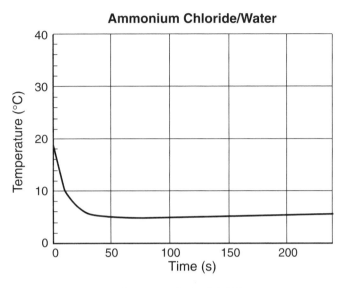

Calcium Chloride/Water

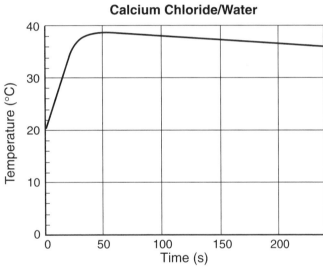

Sodium Bicarbonate/Hydrochloric Acid

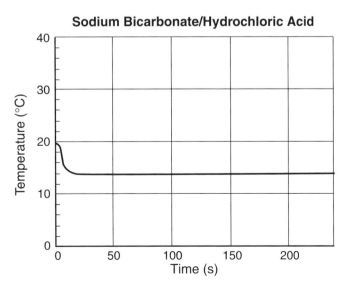

Answers to Post-Lab Questions *(Student answers will vary.)*

1. Complete the following Results Table to indicate whether each reaction in Part A represents a physical or chemical change and whether it is exothermic or endothermic.

Reaction	Physical or Chemical Change?	Exothermic or Endothermic?
A	Physical change	Endothermic
B	Physical change	Exothermic
C	Chemical change	Endothermic

2. A chemical change involves a change in the composition of matter—the formation of a new chemical substance (product) with physical and chemical properties different from those of the reactants. Describe the evidence used to decide if any of the processes in Part A were chemical changes.

   Reaction C—sodium bicarbonate and hydrochloric acid—produced a new chemical substance, one that is a gas at room temperature (or slightly lower, because the reaction container felt cold to the touch). None of the reactants is a gas at room temperature.

3. Did you observe any qualitative differences in the amount of heat generated in the reactions that were characterized as endothermic in Part A?

   There were two endothermic reactions—dissolving ammonium chloride and the reaction of sodium bicarbonate with hydrochloric acid. Of these two reactions that felt cold to the touch, the ammonium chloride reaction was definitely colder than the bicarbonate reaction. **Note to teacher:** *The quantities involved in these two reactions are not really comparable.*

4. Consider Reaction A: Was energy released or absorbed by the reactants in this system? When you *touched* the reaction container (the plastic bag) was energy being released or absorbed by your *hand*?

   Reaction A was an endothermic reaction—the bag felt cold to the touch. The reactants and products absorbed heat from the surroundings, so the temperature of the surroundings decreased. When I touched the plastic bag, heat energy was being removed from my hand (the hotter object) and transferred to the bag (the colder object). Heat energy is always transferred from an object at a higher temperature to an object at a lower temperature.

5. Write a balanced equation for each of the processes in Part A. Remember to include heat on the reactant or product side, as appropriate.

   Reaction A: $NH_4Cl(s)$ + heat $\rightarrow$ $NH_4Cl(aq)$

   Reaction B: $CaCl_2(s)$ $\rightarrow$ $CaCl_2(aq)$ + heat

   Reaction C: $NaHCO_3(s)$ + $HCl(aq)$ + heat $\rightarrow$ $NaCl(aq)$ + $H_2O(l)$ + $CO_2(g)$

Students may need some gentle reminders about the products formed in Reaction C. Ask students to think about the gas that was observed and its possible identity.

6. In Part B, was the temperature that was measured a part of the system or the surroundings? How long did it take for the maximum or minimum temperature to be reached?

    In all cases, the temperature that was measured was that of the surroundings (the solvent). The maximum or minimum temperature was reached within 10–40 seconds, depending on the reaction. The fastest temperature change was observed for calcium chloride dissolving in water. **Note to teacher:** *You may want to discuss with students why the temperature change is not instantaneous.*

7. Describe in words the temperature versus time graph that was recorded in Part B. Be as specific as possible.

    The initial part of the temperature versus time graph (0–30 seconds) is almost linear and has a very steep slope—the temperature rapidly increases (or decreases) with time. The maximum (or minimum) temperature is reached within 10–40 seconds. The temperature change then levels off and the temperature remains practically constant with time. After approximately 120 sec the temperature begins to decrease (or increase) very slowly, about 0.1 °C every 10 seconds.

8. Complete the following sentence to summarize the observations and conclusions for the reaction in Part B: The reaction of _____ with _____ is an (exothermic/endothermic) process in which energy was (absorbed/released) by the system and the temperature of the surroundings (increased/decreased).

    Ammonium chloride with water: *endothermic; absorbed; decreased.*

    Calcium chloride with water: *exothermic; released; increased.*

    Sodium bicarbonate with hydrochloric acid: *endothermic; absorbed; decreased.*

Measuring Energy Changes
Heat of Fusion

Introduction

A physical change of state or a chemical reaction may be either exothermic or endothermic. An exothermic reaction releases heat to its surroundings, while an endothermic reaction absorbs heat from its surroundings. What is the source of the heat energy that is released in an exothermic reaction? What happens to the heat energy that is absorbed in an endothermic reaction? Can the amount of heat energy be measured?

Concepts

- Heat vs. temperature
- Exothermic vs. endothermic
- Heat of fusion
- Heat of vaporization
- Enthalpy change

Background

Our everyday experience tells us that energy in the form of heat is needed to melt ice or boil water. Imagine a beaker of water at room temperature on a hot plate. As the beaker is heated, the temperature of the water will increase steadily from 25 °C to 100 °C, the boiling point of water. If the water is heated further, it will begin to boil, but the temperature of the water in the beaker will remain constant at 100 °C until all of the water has been converted to steam. All of this time, of course, while the temperature has remained constant, heat has been added to the system. The heat absorbed by the water has been used to break apart the forces (e.g., hydrogen bonding) between water molecules in the liquid phase. The amount of heat that must be absorbed to vaporize a specific quantity of liquid (usually one gram or one mole) is called the *heat of vaporization*. The heat of vaporization for water is 2260 Joules per gram. In a similar manner, heat is also required to melt ice. The amount of heat that must be absorbed to melt a specific quantity of solid is called the *heat of fusion*.

Vaporization and fusion are examples of endothermic physical changes. The reverse physical processes must therefore be exothermic. When steam condenses to water, for example, an amount of heat energy equal to the heat of vaporization must be released to the surroundings. This explains why steam burns are more dangerous than hot water burns. Heat is also released when water freezes to ice.

The amount of heat transferred in these processes depends on the difference in the energy stored in each substance. This stored energy is called the heat content or enthalpy, and is represented by the symbol H. The *enthalpy change* (ΔH) for a physical process or a chemical reaction is defined as the heat change that occurs at constant pressure. This is convenient, because most of the reactions that are carried out in the lab are in flasks or containers that are open to the atmosphere—that is, they take place at a constant pressure equal to the barometric pressure.

The temperature description for the beaker of boiling water refers to the "normal" boiling point of water. The boiling point of any liquid depends on pressure—the normal boiling point is the boiling point at 1 atm pressure.

Equation 1 shows the equality between the change in enthalpy (ΔH) of a system and the amount of heat transferred, symbolized by q_P, for a reaction carried out at constant pressure.

$$\Delta H = q_P \qquad \textit{Equation 1}$$

The amount of heat (q_P) transferred to a substance or object depends on three factors: the mass (m) of the object, its specific heat (s), and the resulting temperature change (ΔT). See Equation 2.

$$q_P = m \times s \times \Delta T \qquad \textit{Equation 2}$$

The specific heat (s) of a substance reflects its ability to absorb heat energy and is defined as the amount of heat needed to raise the temperature of one gram of a substance by one degree Celsius. The specific heat of water is equal to 4.18 J/g·°C. The temperature change (ΔT) is equal to the difference between the final temperature and the initial temperature ($\Delta T = T_{final} - T_{initial}$).

In most laboratory situations, the temperature change is measured not for the system itself (the reactants and products), but for the surroundings (the solution and the reaction vessel). The amount of heat released by the system must be equal to the amount of heat absorbed by the surroundings. The sign convention in Equation 3 reveals that the heat change occurs in the opposite direction.

$$q(\text{system}) = -q(\text{surroundings}) \qquad \textit{Equation 3}$$

For an exothermic reaction, the heat released by the system results in a temperature increase for the surroundings (ΔT is positive) and the heat absorbed by the surroundings will be a positive quantity. The heat released by the system must have the reverse sign—it must be a negative quantity. According to this convention, the enthalpy change for an exothermic reaction is always a negative value. For an endothermic reaction, in contrast, the heat absorbed by the system results in a temperature decrease for the surroundings (ΔT is negative) and the heat released by the surroundings will be a negative quantity. The heat absorbed by the system must have the opposite sign—it must be a positive quantity. According to this convention, the enthalpy change for an endothermic reaction is always a positive value.

Experiment Overview

The purpose of this experiment is to determine the temperature and heat changes that occur when ice melts. In Part A a heating curve will be obtained by melting ice in a beaker on a hot plate and measuring temperature as a function of time. In Part B ice will be placed in a beaker of warm water and the temperature change that is produced as the ice melts will be measured and used to calculate the heat of fusion of water (the heat required to melt ice).

Refer students to their textbooks for a complete description of enthalpy and heat of reaction. For extra credit, ask students to construct a table summarizing the meanings of the prefixes and the signs of ΔT and ΔH for both exothermic and endothermic reactions.

Teacher Notes

Pre-Lab Questions

A sample of lauric acid—an organic compound used to make soap—was heated above its melting point in a test tube and then cooled in a bath of water until it solidified. The temperature of the lauric acid was measured as a function of time and the following graph was recorded.

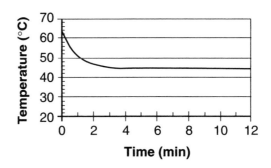

1. What is the physical significance of the flat region (plateau) on the curve?

2. Use the graph to estimate the melting point of lauric acid.

3. Is heat being absorbed or released by the lauric acid sample as it solidifies?

Materials

Beakers, 400-mL, 2	Digital thermometer
Distilled or deionized water	Hot plate
Graduated cylinders, 100- and 250-mL	Insulated foam (Styrofoam™) cup, 6 oz
Ice, crushed, about 200 mL volume	Stirring rod
Ice cubes, 2	Beaker tongs or Hothands®

Safety Precautions

Exercise caution when using the hot plate and handling hot glassware. Remember that a "hot" hot plate looks exactly the same as a "cold" hot plate. Use beaker tongs or Hothands® to handle hot glassware. Wear chemical splash goggles whenever chemicals, heat, or glassware are used in the laboratory.

Procedure

Part A. Temperature and Phase Changes

1. Obtain a 400-mL beaker and fill the beaker to the 150-mL mark with crushed ice.

2. Place the digital thermometer in the ice and measure its temperature (it should be between 0 and 2 °C). Record the lowest temperature as the initial temperature (time 0) in Data Table A.

3. Place the beaker on the hot plate and adjust the heat setting to the halfway point (for example, if the heat dial goes from 0–10, adjust the setting to 5). Stir the ice constantly using a stirring rod.

Make sure the beakers used in Part A are made of Pyrex® and are free of chips and cracks.

4. Measure and record the temperature of the ice every minute. Note in Data Table A the temperature at which all of the ice has melted.

5. When all of the ice has melted, adjust the hot plate setting to its maximum value (10 in the above example) and continue heating and stirring the water. Do not allow the thermometer to touch the bottom of the beaker.

6. Record the temperature of the water every minute. Note in Data Table A the temperature at which the water begins to boil. Continue heating until the water has boiled for three minutes.

7. Turn off the hot plate and carefully remove the beaker using beaker tongs or Hothand® heat protectors.

Part B. Energy Needed to Melt Ice

8. Obtain 100 mL of *warm* water (about 50 °C) in a graduated cylinder. Measure the initial volume of water to the nearest 0.2 mL and record this value in Data Table B.

9. Place a Styrofoam cup in a 400-mL beaker for stability and pour the water into the cup. Measure and record the temperature of the warm water to the nearest 0.1 °C in Data Table B.

10. Obtain several ice cubes. Shake any excess water off the ice cubes and carefully add the ice cubes to the warm water bath.

11. Stir the ice/water mixture until the temperature is around 0 °C (within 2–4 °C). Add more ice, if necessary, to cool the water to this temperature.

12. When the temperature has reached its lowest value (again, this should be between 0 and 4 °C), record the temperature and immediately remove any unmelted ice from the water bath using tongs.

13. Carefully pour the water from the Styrofoam cup into a 250-mL graduated cylinder. Measure and record the final volume of water to the nearest 1 mL in Data Table B. *Note:* If a 250-mL graduated cylinder is not available, measure the volume of water in two batches using a 100-mL graduated cylinder.

Hot tap water is a convenient source of warm water for Part B. The temperature of the water should be between 40 and 50 °C for best results.

Teacher Notes

Name: _____

Class/Lab Period: _____

Measuring Energy Changes

Data Table A. *Temperature and Phase Changes*

Time (min)	Temperature (°C)	Notes	Time (min) continued	Temperature (°C) continued	Notes
0 (initial)					

Data Table B. *Energy Needed to Melt Ice*

Initial Volume (Warm Water)	Initial Temperature (Warm Water)	Final Volume (Ice Water)	Final Temperature (Ice Water)

Post-Lab Calculations and Analysis *(Use a separate sheet of paper to answer the following questions.)*

Part A. Temperature and Phase Changes

1. Using your data, draw a graph of temperature (y-axis) versus time (x-axis).

2. Note on the graph the temperature at which all of the ice has melted and the temperature at which the water starts to boil.

3. Are there any temperature plateaus (flat regions of the curve) on the temperature versus time graph? Name the physical property of water corresponding to each temperature plateau.

4. Was heat being added to the system during the times that the temperature remained relatively constant? Use the law of conservation of energy to describe what happened to the heat energy that was absorbed during this time.

5. Was heat being added to the system during the times that the temperature was rising? Use the law of conservation of energy to describe what happened to the heat energy that was absorbed during this time.

Part B. Energy Needed to Melt Ice

6. Use the appropriate density value from the following table of density versus temperature to calculate the mass of warm water that was cooled by the ice in Part B.

Temperature	40 °C	45 °C	50 °C	55 °C
Density of water	0.992 g/mL	0.990 g/mL	0.988 g/mL	0.985 g/mL

7. Use Equation 2 in the *Background* section to calculate the amount of heat in joules released by the warm water as it cooled.

8. Use Equation 3 to calculate the amount of heat absorbed by the ice as it melted.

9. Determine the volume of ice melted in Part B and calculate its mass. *Note:* The density of water at 0 °C is 1.00 g/mL. Divide the result in Question #8 by the mass of ice melted to determine the amount of energy absorbed per gram of ice as it melted.

10. (a) Use the gram formula weight of water to calculate the heat of fusion of water in *kilojoules per mole* (kJ/mole).

    (b) Write a chemical equation for melting ice and include the energy term in kJ/mole.

    (c) The literature value for the heat of fusion of water is 6.02 kJ/mole. Calculate the percent error in the experimentally determined heat of fusion.

$$\text{Percent error} = \frac{|\text{ experimental } - \text{ literature }|}{\text{literature}} \times 100\%$$

Teacher's Notes
Measuring Energy Changes

Master Materials List *(for a class of 30 students working in pairs)*

Beakers, 400-mL, 30	Digital thermometers or temperature sensors, 15†
Distilled or deionized water	Hot plates, 8*
Graduated cylinders, 100-mL, 15	Insulated foam (Styrofoam®) cups, 6 oz, 15
Graduated cylinders, 250-mL, 15	Stirring rods, 15
Ice, about 3 L volume	Beaker tongs or Hothands®, 15

*Two groups may share a large hot plate. Stagger the starting points for Parts A and B and adjust the number of students in each group to accommodate the number of hot plates available.

†See the *Supplementary Information* section for directions on how to use technology for data collection.

Safety Precautions

Exercise caution when using the hot plate and handling hot glassware. Remember that a "hot" hot plate looks exactly the same as a "cold" hot plate. Use beaker tongs or Hothands® to handle hot glassware. Wear chemical splash goggles whenever chemicals, heat, or glassware are used in the laboratory.

Disposal

Water may be poured down the drain.

Lab Hints

- The experimental work for this laboratory can reasonably be completed in one 50-minute lab period. Part A takes about 25 minutes. *Note:* Remind students to increase the setting on the hot plate to its maximum value as soon as the ice has melted (otherwise Part A will take more than 30 minutes). Part B takes 20 minutes to complete.

- The number of hot plates is probably the "limiting reagent" for this lab. Stagger the starting points for Parts A and B and adjust the size of the student groups to make this experiment work in your lab setting. *For example:* Five hot plates would be needed for a class of 30 students working in groups of three, where half the groups start with Part A and half with Part B. Also, if large (7″ × 7″) hot plates are used, two groups of students may share one hot plate.

- If too many hot plates are being used simultaneously, they may trip a circuit breaker.

- Consider pooling the class data for Part B to calculate the heat of fusion of ice (Question #10, Post-Lab Calculations and Analysis). Use the class average for the amount of heat absorbed per gram of ice to calculate the heat of fusion of water in kilojoules per mole and to analyze the percent error in the experiment.

- See the *Supplementary Information* section for directions on how to adapt the procedure to the use of technology for data collection and analysis.

Teaching Tips

- It may be helpful to perform a sample heat energy calculation exercise with the class prior to beginning this activity. The "Specific Heat" demonstration in this Flinn ChemTopic™ Labs manual provides excellent data to illustrate the nature of the heat equation (Equation 2) and perform sample heat calculations.

- Enthalpy is an abstract concept that is often difficult for students to understand. Most textbooks include diagrams of enthalpy versus "reaction coordinate" (reactants and products) that help students visualize the difference in the sign of ΔH for exothermic and endothermic reactions. Laboratory activities such as "Exploring Energy Changes" and "Discovering Instant Cold Packs" are very helpful in teaching enthalpy because they allow students to see and feel the real effects of enthalpy changes.

Answers to Pre-Lab Questions *(Student answers will vary.)*

A sample of lauric acid—an organic compound used to make soap—was heated above its melting point in a test tube and then cooled in a bath of water until it solidified. The temperature of the lauric acid was measured as a function of time and the following graph was recorded.

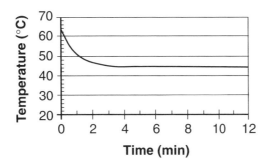

1. What is the physical significance of the flat region (plateau) on the curve?

    The flat region of the curve corresponds to the temperature at which the molten (liquid) lauric acid solidifies. At this point on the cooling curve, the lauric acid sample is still releasing energy to its surroundings–the source of the energy is that released by the molecules as they get closer together in the solid phase than in the liquid phase and thus have greater intermolecular forces between them.

2. Use the graph to estimate the melting point of lauric acid.

    The melting point of lauric acid is approximately 44 °C.

3. Is heat being absorbed or released by the lauric acid sample as it solidifies?

    Heat is released as lauric acid molecules in the liquid phase condense to form a solid. This is an exothermic process. The source of the heat energy is the energy released when stronger intermolecular bonds are formed between molecules in the more closely packed solid phase compared to the liquid phase.

Sample Data

Student data will vary.

Data Table A. *Temperature and Phase Changes*

Time (min)	Temperature (°C)	Notes	Time (min) continued	Temperature (°C) continued	Notes
0 (initial)	0 °C	Solid ice.	14	32.0	
1	0.3		15	40.6	
2	0.4		16	48.9	
3	0.7	Ice begins to melt.	17	56.0	
4	1.0		18	64.9	
5	1.0		19	72.3	
6	1.0		20	80.0	
7	1.2		21	88.3	
8	2.0		22	96.0	Water begins to boil.
9	3.8		23	101.1	
10	6.3	Ice has completely melted.	24	101.0	
11	10.1		25	101.0	
12	16.1		26	100.9	
13	24.5		27	100.9	Water and steam.

Data Table B. *Energy Needed to Melt Ice*

Initial Volume (Warm Water)	Initial Temperature (Warm Water)	Final Volume (Ice Water)	Final Temperature (Ice Water)
100.0 mL	46.8 °C	154 mL	1.1 °C

Possible Answers to Post-Lab Calculations and Analysis *(Student answers will vary.)*

Part A. Temperature and Phase Changes

1. Using your data, draw a graph of temperature (y-axis) versus time (x-axis).

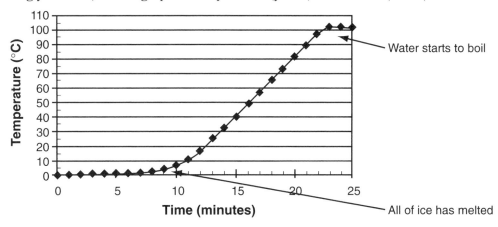

2. Note on the graph the temperature at which all of the ice has melted and the temperature at which the water starts to boil.

   See Sample Graph.

3. Are there any temperature plateaus (flat regions of the curve) on the temperature versus time graph? Name the physical property of water corresponding to each temperature plateau.

   There are two temperature plateaus on the temperature versus time graph. The first temperature plateau occurs at approximately 0–2 °C and corresponds to the melting point of ice. A second temperature plateau occurs at about 100–101 °C and corresponds to the boiling point of water.

4. Was heat being added to the system during the times that the temperature remained relatively constant? Use the law of conservation of energy to describe what happened to the heat energy that was absorbed during this time.

   Heat energy was being added to the system while the temperature remained constant between 0 and 2 °C. The added heat energy energy disrupted the forces holding the water molecules in a close and orderly arrangement in the solid phase. Heat was also being added to the system while the temperature remained constant between 100 and 101 °C. The added heat energy broke the intermolecular attractive forces holding the water molecules together in the liquid phase.

5. Was heat being added to the system during the times that the temperature was rising? Use the law of conservation of energy to describe what happened to the heat energy that was absorbed during this time.

   Heat energy was continuously being added to the system while the temperature was rising in a linear fashion between approximately 5 and 95 °C. The added heat energy was converted to increased kinetic energy of the water molecules in the liquid phase. The average speed of the molecules increased. The average kinetic energy of molecules is directly proportional to their absolute temperature in degrees Kelvin.

Teacher Notes

Part B. Energy Needed to Melt Ice

6. Use the appropriate density value from the following table of density versus temperature to calculate the mass of warm water that was cooled by the ice in Part B.

Temperature	40 °C	45 °C	50 °C	55 °C
Density of water	0.992 g/mL	0.990 g/mL	0.988 g/mL	0.985 g/mL

The density of water at 48 °C is approximately 0.99 g/mL. The mass of warm water is equal to 100.0 mL × 0.99 g/mL = 99 g.

7. Use Equation 2 in the *Background* section to calculate the amount of heat in joules released by the warm water as it cooled.

$$\Delta T = T_{final} - T_{initial} = 1.1 - 46.8\ °C = -45.7\ °C$$

$$q(warm\ water) = (4.18\ J/g{\cdot}°C) \times 99\ g \times (-45.7\ °C) = -18,900\ J$$

8. Use Equation 3 to calculate the amount of heat absorbed by the ice as it melted.

$$q(ice) = -q(warm\ water) = +18,900\ J$$

9. Determine the volume of ice melted in Part B and calculate its mass. *Note:* The density of water at 0 °C is 1.00 g/mL. Divide the result in Question #8 by the mass of ice melted to determine the amount of energy absorbed per gram of ice as it melted.

$$Volume\ of\ ice\ melted = Final\ volume - Initial\ volume = 154 - 100\ mL = 54\ mL$$

$$Mass\ of\ ice\ melted = 54\ mL \times 1.00\ g/mL = 54\ g$$

Amount of energy absorbed per gram of ice as it melted = q(ice)/mass of ice =
18,900 J/54 g = 350 J/g

10. (a) Use the gram formula weight of water to calculate the heat of fusion of water in *kilo-joules per mole* (kJ/mole).

    (b) Write a chemical equation for melting ice and include the energy term in kJ/mole.

    (c) The literature value for the heat of fusion of water is 6.02 kJ/mole. Calculate the percent error in the experimentally determined heat of fusion.

$$Percent\ error = \frac{|\ experimental - literature\ |}{literature} \times 100\%$$

(a) Heat of fusion (kJ/mole) = 350 J/g × 18 g/mole × 1 kJ/1000 J = 6.3 kJ/mole

(b) $H_2O(s) + 6.3\ kJ/mole \rightarrow H_2O(l)$

(c) Percent error = $\dfrac{|\ 6.3 - 6.02\ |}{6.02} \times 100\% = 5\%$

Supplementary Information

The following instructions are provided for adapting Part A to the use of technology (calculator- or computer-interfaced data collection and analysis).

1. Connect the interface (LabPro, CBL system, etc.) to a computer or calculator.

2. Plug a temperature sensor into the interface.

3. Open and set up a graph in the data collection software so that the y-axis reads temperature in degrees Celsius. Set the minimum and maximum temperature values at –10 degrees and 110 degrees, respectively.

4. The x-axis should be set for time in minutes. Set the minimum and maximum time values at 0 minutes and 25 minutes.

5. The time interval should be set so a reading is taken every minute.

6. Obtain 150 mL of crushed ice in a 400-mL beaker and place the temperature sensor in the ice for a few seconds to allow the sensor to become acclimated to the initial temperature.

7. Place the beaker on the hot plate and adjust the heat setting to the halfway point. Stir the ice constantly using a stirring rod.

8. Press Start to begin collecting temperature data.

9. When the ice has melted, increase the hot plate setting to its maximum value.

10. Continue collecting temperature data for 25 minutes.

11. The LabPro or CBL-2 interface will automatically stop collecting data at the preset maximum time value.

12. Use the data in the data table and graph to answer the questions in Part A of the PostLab Calculations and Analysis.

Visit the Vernier website at www.vernier.com to download experiment details.

Discovering Instant Cold Packs
Heat of Solution

Teacher Notes

Introduction

Instant cold packs are familiar first aid devices used to treat injuries when ice is unavailable. Most commercial cold packs consist of a plastic package containing a white solid and an inner pouch of water. Firmly squeezing the pack causes the inner pouch to break. The solid then dissolves in the water producing a change in temperature. Can we measure the temperature change that occurs when the cold pack solid dissolves in water and determine the heat change for this process?

Concepts

- Enthalpy change
- Calorimetry
- Heat of solution
- Dependent and independent variables

Background

The energy or enthalpy change associated with the process of a solute dissolving in a solvent is called the heat of solution (ΔH_{soln}). In the case of an ionic compound dissolving in water, the overall energy change is the net result of two processes—the energy required to break the attractive forces (ionic bonds) between the ions in the crystal lattice, and the energy released when the dissociated (free) ions form ion-dipole attractive forces with the water molecules.

Heats of solution and other enthalpy changes are generally measured in an insulated vessel called a *calorimeter* that reduces or prevents heat loss to the atmosphere outside the reaction vessel. The process of a solute dissolving in water may either release heat into the aqueous solution or absorb heat from the solution, but the amount of heat exchange between the calorimeter and the outside surroundings should be minimal. When using a calorimeter, the reagents being studied are mixed directly in the calorimeter and the temperature is recorded both before and after the reaction has occurred. The amount of heat change occurring in the calorimeter may be calculated using the following equation: $q = m \times s \times \Delta T$, where m is the total mass of the solution (solute plus solvent), s is the specific heat of the solution, and ΔT is the observed temperature change. The specific heat of the solution is generally assumed to be the same as that of water, namely, 4.18 J/g·°C.

Experiment Overview

The purpose of this inquiry-based experiment is to design and carry out a procedure to determine the enthalpy change that occurs when a "cold pack solid" dissolves in water.

Pre-Lab Questions

Consider the following questions or guidelines:

1. What information (data) is needed to calculate an enthalpy change for a reaction?
2. Identify the variables that will influence the experimental data.
3. What variables should be controlled (kept constant during the procedure)?

Students should be familiar with the following definitions before beginning this activity: exothermic and endothermic reactions, heat and temperature, and the system versus the surroundings.

4. The independent variable in an experiment is the variable that is changed by the experimenter, while the dependent variable responds to (depends on) changes in the independent variable. Choose the dependent and independent variables for this experiment.

5. Discuss the factors that will affect the precision of the experimental results.

Materials

Beaker, 400-mL	Balance, centigram (0.01 g precision)
"Cold pack solid," 15 g	Digital thermometer or temperature sensor
Distilled or deionized water	Spatula
Graduated cylinder, 100-mL	Stirring rod
Insulated foam (Styrofoam™) cups, 6 oz	Weighing dishes

Safety Precautions

The cold pack solid is slightly toxic by ingestion and is a body tissue irritant. Avoid contact of all chemicals with eyes and skin. Wear chemical splash goggles and chemical-resistant gloves and apron. Wash hands thoroughly with soap and water before leaving the laboratory.

Procedure

Part A. What Is an Instant Cold Pack?

Complete the following activity to become familiar with the nature and amounts of materials in a commercial cold pack.

1. Obtain a label of a commercial cold pack and write the name of the solid used in the pack.

2. Read the warning information on the label and record any hazards associated with this product.

3. Using the known charges of ions, write the formula of the solid.

4. Calculate the molar mass of the solid.

5. Determine the total mass of the solid: Tare a large weighing dish or cup on the balance. Transfer the cold pack solid to the tared weighing dish. Record the mass of the solid to the nearest 0.01 g.

6. Calculate the number of moles of solid in the pack.

7. Measure the volume of water contained in the inner pouch.

8. Calculate the mass of water in the commercial cold pack (assume the density of water is 1.0 g/mL).

Name of solid	
Warning	
Formula of solid	
Molar mass	
Mass of solid	
Moles of solid	
Volume of water	
Mass of water	

See the Lab Hints *and* Supplementary Information *sections for information about using technology for data collection and analysis.*

Teacher Notes

Part B. Measuring the Heat of Solution

Design and carry out a procedure to determine the enthalpy change (ΔH_{soln}) that occurs when the cold pack solid dissolves in water. Use a maximum of 5 grams of solid per measurement. Write out the procedure in steps and construct a data table that clearly shows the data that will be collected and the measurements that will be made. Have your teacher check the procedure and data table before beginning the experiment.

Procedure

1.

2.

3.

4.

5.

6.

Data Table. Enthalpy Change for Dissolving the Cold Pack Solid

Name: _____

Class/Lab Period: _____

Discovering Instant Cold Packs

Post-Lab Calculations and Analysis

1. Calculate the *heat energy change in joules* when the cold pack solid dissolved in water in your experiment. *Recall:* $q = m \times s \times \Delta T$, where s (specific heat of water) is equal to 4.18 J/g·°C.

2. Calculate the energy change in *joules per gram of solid* for the cold pack solid dissolving in water.

3. Calculate the energy change in units of *kilojoules per mole* of solid for the cold pack solid dissolving in water. To do this:

   (a) Convert the heat energy change found in Question #1 to kilojoules.

   (b) Convert the grams of solid used in the experiment to moles.

   (c) Divide the energy change in kilojoules by the number of moles of solid to determine the energy change in units of kJ/mole. If more than one trial was performed, calculate the *average value* of the heat of solution also.

4. Using the result from Question #3c and the information obtained in Part A, calculate the number of kilojoules involved when the entire cold pack is activated.

5. Circle the correct choices in the following sentence to summarize the heat change that occurs when the commercial cold pack is activated:

   "When the white solid in the commercial cold pack dissolves in water, the pack feels (*hot/cold*) because the temperature of the solution (*increases/decreases*). Energy is (*absorbed/released*) from the surroundings during this reaction and the reaction is classified as (*endothermic/exothermic*). The sign of ΔH for the heat of solution is (*positive/negative*)."

To increase the level of student independence in this inquiry-based experiment, do not give the students these directions. Let them decide on their own how to calculate the enthalpy change for the process.

Teacher's Notes
Discovering Instant Cold Packs

Master Materials List *(for a class of 30 students working in pairs)*

Beakers, 400-mL, 15

"Cold pack solid," Ammonium Nitrate, NH_4NO_3, 150–225 g*

Distilled or deionized water, 1–2 L

Graduated cylinders, 100-mL, 15

Insulated foam (Styrofoam™) cups, 15–30

Balances, centigram (0.01 g precision), 3

Digital thermometers or temperature sensors, 15

Spatulas, 15

Stirring rods, 15

Weighing dishes, 15

*The amount of solid required depends on the number of runs or trials that each group does. A minimum of 2–3 runs is recommended to average the effects of random error. One instant cold pack contains about 185 g, enough for two runs by each group.

Safety Precautions

The cold pack solid is ammonium nitrate. It is slightly toxic by ingestion and is a body tissue irritant. Ammonium nitrate in solid form is a strong oxidizer and may explode if heated or in contact with combustible materials. Avoid contact of the dry solid with combustible organic materials. Avoid contact of all chemicals with eyes and skin. Wear chemical splash goggles and chemical-resistant gloves and apron. Please consult current Material Safety Data Sheets for additional safety, handling, and disposal information.

Disposal

Please consult your current *Flinn Scientific Catalog/Reference Manual* for general guidelines and specific procedures governing the disposal of laboratory waste. The cold pack solutions generated in this experiment may be rinsed down the drain with excess water according to Flinn Suggested Disposal Method #26b.

Lab Hints

• The laboratory work for this inquiry-based, "discovery" experiment can reasonably be completed in one 50-minute lab period. Part A, obtaining information from the cold pack label, takes about 10 minutes. The actual experimental measurements for Part B are very quick—three independent trials are easily completed within 20 minutes.

• To minimize the amount of cold pack solid required, class data may be collected in lieu of 2–3 runs per group. Averaging the class results will tend to eliminate outlying data.

• The most important element for success in an inquiry-based activity is student preparation. Sufficient time should be alotted for students to think through the measurements that must be made, how they will be made, the variables that will influence the measurements, and how the variables can be controlled, if necessary. The *PreLab* Section provides some leading questions and guidelines to stimulate class discussion. The questions may be used as the basis of a small-group activity prior to lab or assigned as homework in preparation for lab. Encourage students to work together to devise a procedure for the calorimetry experiment.

- Temperature measurements may be made using digital thermometers, glass-bulb thermometers, or computer-interfaced temperature sensors. Digital thermometers are preferred over glass thermometers because they provide direct readings, update every second, and have a precision of ±0.1 °C. Glass thermometers are fragile and easily broken, especially if the solutions are vigorously stirred, as suggested in the Sample Procedure. In addition, the 1 °C divisions that are marked on most glass thermometers make them less precise (±0.5 °C) than digital thermometers. Never allow students to use a glass thermometer as a stirring rod.

- See the *Supplementary Information* section for directions on how to adapt the procedure to use a temperature sensor or probe for computer-interfaced data collection and analysis.

- The minimum temperatures recorded in the *Sample Data* section were generally achieved within one minute after mixing and were usually stable for an additional minute. Both of these factors ensure that temperature is easily and precisely measured and that students will feel confident about their measurements.

- Two Styrofoam cups nested together provide better insulation and thermal stability than one cup. If two cups are used, students can easily run two trials without rinsing and drying the cup between trials. Simply have students interchange the actual solution cup and the bottom cup between measurements. In addition, we recommend that students nestle the Styrofoam cup in a beaker for added stability when the thermometer is placed in the cup.

- The volume of liquid is a variable that must be controlled in the experiment, but there is no obvious or absolute value that should be used. The minimum volume of water needed to ensure that the thermometer is completely immersed in liquid is about 20–30 mL. This rules out using a volume of water proportional to the amount of water in a commercial cold pack (which has approximately 185 g of solid and only 80 mL of water). If too large a liquid volume is used, the observed temperature change will be small and may be less reliable. Working with 5 g of cold pack solid and a water volume of 30 mL gave three significant figures in the measured temperature difference. It should be noted, however, that excellent results were also obtained when the reaction was carried out with 100 mL of distilled water.

Teaching Tips

- One of the more stubborn student misconceptions is the idea that if the reaction mixture gets cold, it must have lost heat, therefore the reaction must be exothermic. This misconception may be traced to a lack of understanding of the system versus the surroundings. The temperature change that is measured in a typical coffee-cup calorimeter experiment is that of the surroundings. A heat of solution experiment is probably more confusing on this point than a heat of neutralization or heat of combustion experiment, because water is involved in the reaction. Also, using the combined mass of the solute and the solvent in the heat equation to calculate the heat change tends to blur the traditional distinction between the reactants and products versus the solvent.

- Students may compare their experimental results with the literature value for the heat of solution of ammonium nitrate. Assuming a literature value of 25.7 kJ/mole (*CRC Handbook of Chemistry and Physics,* 82nd Edition, Sectiion 5, pg. 105), the results in the *Sample Data* section (26.4 kJ/mole) give a calculated percent error of 3%.

- This Flinn ChemTopic™ Labs manual includes several valuable demonstrations to compare and contrast key concepts that are essential to a thorough understanding of calorimetry. The "Colorful Heat" demonstration distinguishes between heat and temperature. The "Specific Heat of Metals" demonstration highlights both the definition and application of specific heat and illustrates the variables that are involved in the heat equation.

Answers to Pre-Lab Questions *(Student answers will vary.)*

Consider the following questions and guidelines:

1. What information (data) is needed to calculate an enthalpy change for a reaction?

   In order to calculate the enthalpy change for a reaction, data for the three terms involved in the heat energy equation ($q = m \times s \times \Delta T$) must be known or measured. The mass (m) is the mass of the solution after the cold pack solid has dissolved. The specific heat capacity (s) is assumed to be the same as the specific heat capacity of water (4.18 J/g·°C). The temperature change (ΔT) is equal to the difference between the final and initial temperatures ($T_{final} - T_{initial}$).
   Note to teachers: *Assuming the specific heat capacity of the solution is the same as that of water may be a major source of error in the heat calculations.*

2. Identify the variables that will influence the experimental data.

   Some of the critical variables include: (1) the mass of the solute (cold pack solid); (2) the volume (mass) of the solvent; (3) whether all of the solute dissolves in the solvent; (4) the heat insulating properties of the reaction container; (5) how well the reaction mixture is stirred; (6) how stable the initial temperature reading is.

3. What variables should be controlled (kept constant during the procedure)?

   The following variables should be held constant during the procedure: the volume (mass) of the solvent; the type of reaction container that is used (two insulating foam cups nestled one inside the other will provide better insulation than one cup); continuous stirring of the reaction mixture.

4. The independent variable in an experiment is the variable that is changed by the experimenter, while the dependent variable responds to (depends on) changes in the independent variable. Choose the dependent and independent variable for this experiment.

    In a calorimetry experiment, the mass of the solute in grams is the independent variable and will be varied in different trials. The temperature change that is produced depends on the mass of the solute and is thus the dependent variable in a calorimetry experiment.

5. Discuss the factors that will affect the precision of the experimental results.

    Many factors will influence the precision of the results:

    - *The precision of the balance used to measure the mass of solute.*

    - *The precision of the graduated cylinder used to measure the volume of solvent.*

    - *The precision of the thermometer used to measure the temperature of the reaction mixture.*

    - *The number of times the experiment is repeated to average the effects of random errors.*

    - *The type of vessel that is used as the calorimeter—how much heat is gained or lost by the calorimeter itself.*

    The first three measurements should be made with the most precise glassware and equipment available in the lab—centigram balances (at least), appropriate size graduated cylinders, and digital thermometers, if possible. One important way to improve the precision of the experimental results is to average data obtained over several runs or trials. A minimum of 2–3 trials is recommended. Alternatively, class data may be averaged to eliminate outlying results.

Teacher Notes

Sample Data

Student data will vary.

Part A. What Is an Instant Cold Pack?

The following information was obtained using a Thera-Med brand Instant Cold Pack, manufactured by Thera-Med, Inc., Texas.

Name of solid	Ammonium Nitrate
Warning	Use special care with people whose skin may be sensitive to temperature extremes. May cause discomfort.
Formula of solid	NH_4NO_3
Molar mass	80.04 g/mole
Mass of solid	184.48 g
Moles of solid	184.48 g/80.04 g/mole = 2.305 moles
Volume of water	80.0 mL
Mass of water	80.0 g

Part B. Measuring the Heat of Solution

Sample Procedure

1. Set an insulating foam cup in a 400-mL beaker for stability.

2. Measure 50.0 mL of distilled water in a graduated cylinder and add it to the insulating foam cup. Record the volume of water used in the data table.

3. Place a digital thermometer in the water in the cup and allow the thermometer and the water to adjust to room temperature for at least 2–3 minutes.

4. Tare a weighing dish on the balance. Add about 5 g of cold pack solid to the weighing dish. Measure and record the exact mass of solid to the nearest 0.01 g.

5. Measure the initial temperature of the water in the cup to the nearest 0.1 °C and record the value in the data table.

6. Stir the water in the cup with a stirring rod and add the massed amount of cold pack solid to the water. Make sure that all of the solid dissolves in the water.

7. Stir constantly and monitor the temperature of the solution. Record the lowest temperature that is reached.

8. Rinse the cup contents down the drain with plenty of excess water and dry the inside of the cup with a paper towel.

9. Repeat steps 1–8 at least once more. Record all data.

Sample Data Table

Trial	Mass of Cold Pack Solid (g)	Volume of Water (mL)	Mass of Solution (g)	Initial Temperature (°C)	Final Temperature (°C)	Temperature Change, ΔT (°C)
1	5.37	30.0	35.37	19.9	8.1	–11.8
2	4.78	30.0	34.78	19.8	8.8	–11.0
3	4.89	30.0	34.89	20.0	8.9	–11.1

Answers to Post-Lab Calculations and Analysis *(Student answers will vary.)*

1. Calculate the *heat energy change in joules* when the cold pack solid dissolved in water in your experiment. *Recall:* $q = m \times s \times \Delta T$, where s (specific heat of water) is equal to 4.18 J/g·°C.

   Trial 1. $q = 35.37\,g \times 4.18\,J/g{\cdot}°C \times (-11.8\,°C) = -1740\,J$

   Trial 2. $q = 34.78\,g \times 4.18\,J/g{\cdot}°C \times (-11.0\,°C) = -1600\,J$

   Trial 3. $q = 34.89\,g \times 4.18\,J/g{\cdot}°C \times (-11.1\,°C) = -1620\,J$

   Note to teacher: *The negative sign for the heat change indicates that the solution (the surroundings) lost heat energy to the solute (the system) when the solid dissolved in water. According to the law of conservation of energy, the amount of heat lost by the surroundings must be equal to the amount of energy gained by the system for the reaction to occur. The cold pack solid absorbed energy from the surroundings as it dissolved in water—it is an endothermic process. The heat change for an endothermic process is a positive quantity.*

2. Calculate the energy change in *joules per gram of solid* for the cold pack solid dissolving in water.

   Trial 1. $\Delta H_{soln} = 1740\,J/5.37\,g = 324\,J/g$

   Trial 2. $\Delta H_{soln} = 1600\,J/4.78\,g = 335\,J/g$

   Trial 3. $\Delta H_{soln} = 1620\,J/4.89\,g = 331\,J/g$

   Average value = $(324 + 335 + 331)/3 = 330 \pm 4\,J/g$

   Note to teacher: *The heat of solution of ammonium nitrate dissolving in water is a positive quantity, equal in magnitude but opposite in sign to the heat change calculated in Question #1.*

The most common mistake students make in these calculations is to use the mass of the solid in Question #1. The heat change that is calculated in Question #1 is for the surroundings, which is the final solution after the solid has dissolved.

3. Calculate the energy change in units of *kilojoules per mole* of solid for the cold pack solid dissolving in water. To do this:

   (a) Convert the heat energy change found in Question #1 to kilojoules.

   Trial 1. 1740 J × (1 kJ/1000J) = 1.74 kJ

   Trial 2. 1600 J × (1 kJ/1000J) = 1.60 kJ

   Trial 3. 1620 J × (1 kJ/1000J) = 1.62 kJ

   (b) Convert the grams of solid used in the experiment to moles.

   Trial 1. 5.37 g × (1 mole/80.04 g) = 0.0671 moles

   Trial 2. 4.78 g × (1 mole/80.04 g) = 0.0597 moles

   Trial 3. 4.89 g × (1 mole/80.04 g) = 0.0611 moles

   (c) Divide the energy change in kilojoules by the number of moles of solid to determine the energy change in units of kJ/mole. If more than one trial was performed, calculate the *average value* of the heat of solution also.

   Trial 1. 1.74 kJ/0.0671 moles = 25.9 kJ/mole

   Trial 2. 1.60 kJ/0.0597 moles = 26.8 kJ/mole

   Trial 3. 1.62 kJ/0.0611 moles = 26.5 kJ/mole

   Average value = (25.9 + 26.8 + 26.5)/3 = 26.4 ±0.3 kJ/mole

4. Using the result from Question #3c and the information obtained in Part A, calculate the number of kilojoules involved when the entire cold pack is activated.

   Recall (Part A, Question 6): The number of moles of solid in the instant cold pack is 2.305 moles.

   Amount of heat transfer = 26.4 kJ/mole × 2.305 moles = 60.9 kJ.

   Note to teacher: *Some perspective might be interesting here. This amount of heat transfer is sufficient to cool 1 liter (1 kg) of water from 20 to 5°C.*

5. Circle the correct choices in the following sentence to summarize the heat change that occurs when the commercial cold pack is activated:

   "When the white solid in the commercial cold pack dissolves in water, the pack feels (*hot*/**COLD**) because the temperature of the solution (*increases*/**DECREASES**). Energy is (**ABSORBED**/*released*) from the surroundings during this reaction and the reaction is classified as (**ENDOTHERMIC**/*exothermic*). The sign of ΔH for the heat of solution is (**POSITIVE**/*negative*)."

Supplementary Information

The following instructions are provided for adapting Part B to the use of technology (computer-interfaced data collection and analysis).

1. Connect the interface (LabPro, CBL system, etc.) to a computer or calculator.

2. Plug a temperature sensor into the interface.

3. Open and set up a graph in your data collection software so that the y-axis reads temperature in degrees Celsius. Set the minimum and maximum temperature values at 0 degrees and 25 degrees, respectively.

4. The x-axis should be set for time in minutes. Set the minimum and maximum time values at 0 seconds and 180 seconds, respectively.

5. The time interval should be set so a temperature reading is taken every 10 seconds.

6. Obtain 30 mL of distilled water in a graduated cylinder and carefully transfer the water to an insulated foam cup nested in a 400-mL beaker for stability.

7. Place the temperature sensor in the water. Allow the probe to equilibrate at the initial temperature for 1 minute, then press Start to begin collecting temperature data.

8. Stir the water using a stirring rod and add the pre-weighed amount of cold pack solid.

9. Continue stirring the solution and collecting temperature data for three minutes (180 seconds).

10. The LabPro or CBL-2 interface will automatically stop collecting data at the preset maximum time value.

11. Print the computer-generated data table and graph, if possible, and use the data to complete the Data Table and the Post-Lab Calculations.

Visit the Vernier website at www.vernier.com to download experiment details.

Measuring Calories
Energy Content of Food

Introduction

All human activity requires "burning" food for energy. How much energy is released when food burns in the body? How is the calorie content of food determined? Let's investigate the calorie content of different snack foods, such as popcorn, peanuts, marshmallows, and cheese puffs.

Concepts

- Combustion reaction
- Calorimetry
- Nutritional Calorie
- Calorie content of foods

Background

What does it mean to say that we burn food in our bodies? The digestion and metabolism of food converts the chemical constituents of food to carbon dioxide and water. This is the same overall reaction that occurs when organic molecules—such as carbohydrates, proteins, and fats—are burned in the presence of oxygen. The reaction of an organic compound with oxygen to produce carbon dioxide, water, and heat is called a *combustion reaction*. The chemical equation for the most important reaction in our metabolism, the combustion of glucose, is shown in Equation 1.

$$C_6H_{12}O_6 \; + \; 6O_2 \; \rightarrow \; 6CO_2 \; + \; 6H_2O \; + \; \text{heat} \qquad\qquad \textit{Equation 1}$$

Within our bodies, the energy released by the combustion of food molecules is converted to heat energy (to maintain our constant body temperature), mechanical energy (to move our muscles), and electrical energy (for nerve transmission). The total amount of energy released by the digestion and metabolism of a particular food is referred to as its *calorie content* and is expressed in units of nutritional Calories (note the uppercase C). A *nutritional Calorie,* abbreviated Cal, is equivalent to a unit of energy called a kilocalorie, or 1000 calories (note the lower case c). One calorie is defined as the amount of heat required to raise the temperature of 1 gram of water by 1 °C. (This is also the definition of the specific heat of water.) The calorie content of most prepared foods is listed on their nutritional information labels.

Nutritionists and food scientists measure the calorie content of food by burning the food in a special device called a calorimeter. *Calorimetry* is the measurement of the amount of heat energy produced in a reaction. Calorimetry experiments are carried out by measuring the temperature change in water that is in contact with or surrounds the reactants and products. (The reactants and products together are referred to as the system, the water as the surroundings.)

In a typical calorimetry experiment, the reaction of a known mass of reactant(s) is carried out either directly in or surrounded by a known quantity of water and the temperature increase or decrease in the water surroundings is measured. The temperature change (ΔT) produced in the water is related to the amount of heat energy (q) absorbed or released by the reaction system according to the following equation:

$$q = m \times s \times \Delta T \qquad \qquad \textit{Equation 2}$$

where m is the mass of water, s is the specific heat of water, and ΔT is the observed temperature change. As mentioned above, the specific heat of water—defined as the amount of heat required to increase the temperature of one gram of water by 1 °C—is equal to 1 cal/g·°C.

Experiment Overview

The purpose of this experiment is to determine the amount of heat energy released when different snack foods burn and identify patterns in the calorie or energy content of snack foods.

Pre-Lab Questions

1. A candy bar has a total mass of 2.5 ounces. In a calorimetry experiment, a 1.0-g sample of this candy bar was burned in a calorimeter surrounded by 1000. g of water. The temperature of the water in contact with the burning candy bar was measured and found to increase from an initial temperature of 21.2 °C to a final temperature of 24.3 °C.

   (a) Calculate the amount of heat in *calories* released when the 1.0-g sample burned.

   (b) Convert the heat in calories to nutritional Calories and divide by the mass of the burned sample in grams to obtain the *energy content* (also called fuel value) in units of Calories per gram.

   (c) Multiply this value by the total number of grams in the candy bar to calculate the total *calorie content* of the candy bar in Calories. *Hint:* Convert the mass in ounces to grams.

2. Consult the nutritional labels on two of your favorite snack foods: Report their total calorie content (total Calories) and calculate their fuel value (Calories per gram).

Materials

Balance, centigram (0.01 g precision)	Matches
Calorimeter and lid*	Snack foods (cheese puffs, peanuts, marshmallows, popcorn, etc.), 2 pieces
Erlenmeyer flask with plastic spill-rim collar, 125-mL	Stirring rod
Food holder (cork) and pin	Thermometer or temperature sensor
Graduated cylinder, 50-mL	Water

*Or "soda-can" calorimeter. See the *Alternative Procedure* section.

Safety Precautions

Wear safety goggles whenever working with chemicals, glassware, or heat in the laboratory. Exercise care when handling hot glassware and equipment. Allow the burned snack food sample to cool before handling it. The food-grade items that have been brought into the lab are considered laboratory chemicals and are for laboratory use only. Do not taste or ingest any materials in the chemistry laboratory. Wash hands thoroughly with soap and water before leaving the lab.

Refer students to the procedure on page 42 if they will be using a "soda can" calorimeter. If a soda can is used, the Erlenmeyer flask and spill collar will not be needed.

Teacher Notes

Procedure

Part A. Setting Up the Calorimeter

1. Place the calorimeter upright so that the triangle opening is at the base.

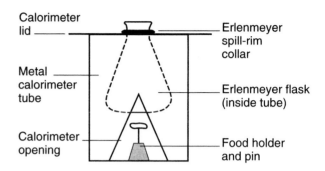

2. Place the mouth of the Erlenmeyer flask through the hole in the calorimeter lid and slide and snap the plastic spill-rim attachment onto the neck of the Erlenmeyer flask.

3. Place the flask and lid assembly onto the top of the calorimeter so that the flask is centered inside the calorimeter.

Part B. The Calorimetry Experiment

4. Measure 50.0 mL of water in a graduated cylinder and pour the water into the Erlenmeyer flask in the calorimeter.

5. Measure the temperature of the water with a thermometer or temperature sensor and record the "initial temperature" of water in the Data Table.

6. Obtain a snack food sample and identify the sample (peanuts, popcorn, etc.) in the Data Table.

7. Place the snack food sample on the food holder and measure the combined mass of the sample and the food holder. Record the "initial mass" of the food sample and holder in the Data Table.

8. Place the food sample and food holder near the triangle opening in the calorimeter and light the food with a match.

9. Quickly slide the food sample and food holder into the base of the calorimeter through the triangle opening. Center the food holder under the Erlenmeyer flask.

10. As the food sample burns, gently stir the water in the calorimeter with a stirring rod. *Note:* Use a stirring rod to stir the water in the calorimeter; never stir with a thermometer.

11. Allow the water to be heated until the food sample stops burning.

12. When the temperature has stabilized, measure the maximum temperature that the water reaches. Record the "final temperature" of the water in the Data Table.

13. Measure and record the "final mass" of the food sample and food holder after it cools.

14. Clean the food holder to remove any traces of food residue.

15. Repeat steps 5–14 with a second food sample.

The food holder and pin consists of a cork stopper, size 7 or 8, and a large straight pin inserted at an angle through the side and top of the cork. For peanuts, use an unfolded paper clip instead of the pin to hold the food item.

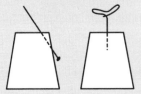

Alternative Procedure. A "Soda-Can" Calorimeter

1. Place a food sample on the food holder. Measure and record the combined mass of the food holder and sample. Place the food holder on a ring stand.

2. Obtain a clean, empty soda can. Measure and record its mass.

3. Add about 50 mL of tap water to the can and measure the combined mass of the can and water.

4. Bend the top tab on the can up and slide a stirring rod through the hole. Suspend the can on a ring stand using a metal ring. Adjust the height of the can so that it is about 2.5 cm above the food holder.

5. Insert a thermometer into the can. Measure and record the initial temperature of the water.

6. Light the food sample and center it under the soda can. Allow the water to be heated until the food sample stops burning. Record the maximum (final) temperature of the water in the can.

7. Measure and record the final mass of the food holder and sample.

8. Clean the bottom of the can and remove any food residue from the food holder. Repeat the procedure with a second food sample.

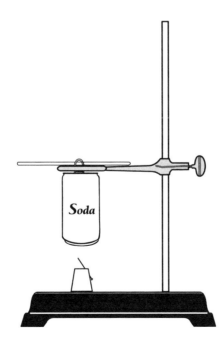

If the tab on the soda can is missing, punch holes through the sides of the soda can and insert the stirring rod through the holes to support the soda can.

Teacher Notes

Name: _____

Class/Lab Period: _____

Measuring Calories

Data Table. *The Calorimetry Experiment*

Food Sample	Initial Mass (Food Sample and Holder), g	Final Mass (Food Sample and Holder), g	Initial Temperature (Water), °C	Final Temperature (Water), °C

Post-Lab Calculations and Analysis *(Use a separate sheet of paper to answer the following questions.)*

Construct a *Results Table* to enter all of the following information and summarize the results.

1. Determine the mass of water heated in the calorimeter for each food sample.

2. Calculate the change in temperature (ΔT) for each sample.

3. Use the heat equation (Equation 2) to calculate the heat (q) absorbed by the water in the calorimeter for each food sample. Report the results in *calories, kilocalories,* and *nutritional Calories.*

4. Subtract the final mass of the food sample and holder from the initial mass to determine the mass in grams of the food sample that burned in each experiment.

5. Use the results from Questions #3 and 4 to calculate the energy content (fuel value) of the food sample in units of *Calories per gram* (Cal/g).

6. Record your results for the energy content of foods along with those of other groups in the class on the overhead projector or the board. Be sure to record the identity of the food sample.

7. Copy all of the results and use the class data to calculate the *average energy content* in units of Cal/g for different types of snack foods. Construct a *Class Results Table* to summarize the results.

8. Rank the snack foods in order of their average energy content, from highest to lowest. Which snack food has the highest energy content? The lowest?

9. Based on your knowledge of the fat content of different snack foods (if necessary, consult their nutritional labels to obtain this information), make a general statement describing the relative energy content of high-fat versus low-fat snack foods.

10. Consider the major sources of error in this experiment. Do you think your results are off on the high side or the low side? Explain.

Teacher's Notes

Measuring Calories

Master Materials List *(for a class of 30 students working in pairs)*

Balance, centigram (0.01 g precision), 3	Matches, 15
Calorimeter and lid, 15*	Snack foods (Cheetos®, marshmallows, peanuts, popcorn, etc.), 30 pieces
Erlenmeyer flask with plastic spill-rim collar, 125-mL, 15*	
	Stirring rods, glass, 15
Food holder and pin, 15*	Thermometers or temperature sensors, 15
Graduated cylinder, 50-mL, 15	Water

*These items are included in the Economy-choice Calorimeter available from Flinn Scientific (Catalog No. AP4533.)

Safety Precautions

Wear safety goggles whenever working with chemicals, glassware, or heat in the laboratory. Exercise care when handling hot glassware and equipment. Allow the burned snack food sample to cool before handling it. The food-grade items that have been brought into the lab are considered laboratory chemicals and are for laboratory use only. Do not taste or ingest any materials in the chemistry laboratory. Burning different foods may generate a large amount of smoke—perform this experiment in a well-ventilated room. Do not use peanuts in this experiment if any students are allergic to peanuts.

Disposal

Please consult your current *Flinn Scientific Catalog/Reference Manual* for general guidelines and specific procedures governing the disposal of laboratory waste. All of the burned food samples should be allowed to cool and then disposed of in the trash according to Flinn Suggested Disposal Method #26a.

Lab Hints

- The laboratory work for this experiment can reasonably be completed in one 50-minute lab period. This should be sufficient time for students to perform two trials, either of the same food or different foods. Carrying out the experiment on two samples of the same food is preferred from the point of view of averaging experimental error, but many students will be curious and want to test different foods. Using the class data to calculate the average energy content of different foods should eliminate some of the fluctuations due to random error.

- Wrap the cork food holder in aluminum foil to prevent the cork from burning. Students should practice sliding the food holder and food assembly into the calorimeter. This should be done quickly after the food has been ignited. The food holder assembly should be centered under the flask and remain upright. For best results, pin the food sample at one end so the sample "points up" and the length is parallel to the pin. The food sample does not have to be completely engulfed in flame before it is placed in the calorimeter. A small flame on the food sample will spread and engulf it over time.

The Economy-choice Calorimeter available from Flinn Scientific (Catalog No. AP4533) provides an easy and inexpensive way to measure the calorie content of solid foods. The product includes the calorimeter tube and lid, Erlenmeyer flask and plastic collar, and the food holder and pin.

Teacher Notes

- The food samples burn with a flame only until they turn to charcoal on the outside. Let the samples smolder for a minute or two inside the calorimeter; the temperature of the water will continue to increase after the flame has been extinguished. Measure the maximum temperature of the water inside the calorimeter flask.

- The burning food sample should be close to but not touching the Erlenmeyer flask in the calorimeter. If the food sample is too close to the bottom of the flask, it may extinguish early due to a lack of oxygen. Black soot will deposit on the bottom of the Erlenmeyer flask in the calorimeter when the food sample burns. This should be wiped off with a little water and a paper towel between trials.

- Avoid snack food samples with a high sugar content. These foods get soft as they burn and may fall off of the pin. Different kinds of nuts—walnuts, pecans, almonds, etc.—are also good choices for this experiment.

Teaching Tips

- The "real world" application in this experiment makes it an effective learning exercise for students who tend to lose interest in purely chemically-oriented applications of enthalpy changes and heats of reaction. (The heat of neutralization of hydrochloric acid and sodium hydroxide may not be a pressing issue for many students. Almost all students have thought about and discussed the calorie content of the foods they eat.)

- This experiment also provides an excellent opportunity to discuss chemical potential energy—the energy stored in compounds, including foods, due to their composition and structure. What is the source of the energy released when food burns? Where did that energy originally come from? Why do different types of molecules release different amounts of heat? How is this related to their structure?

Answers to Pre-Lab Questions *(Student answers will vary.)*

1. A candy bar has a total mass of 2.5 ounces. In a calorimetry experiment, a 1.0-g sample of this candy bar was burned in a calorimeter surrounded by 1000. g of water. The temperature of the water in contact with the burning candy bar was measured and found to increase from an initial temperature of 21.2 °C to a final temperature of 24.3 °C.

(a) Calculate the amount of heat in *calories* released when the 1.0-g sample burned.

(b) Convert the heat in calories to nutritional Calories and divide by the mass of the burned sample in grams to obtain the *energy content* (also called fuel value) in units of Calories per gram.

(c) Multiply this value by the total number of grams in the candy bar to calculate the total *calorie content* of the candy bar in Calories. *Hint:* Convert the mass in ounces to grams.

(a) Heat (q) = m (g) × s (cal/g·°C) × ΔT

m = 1000. g; s = 1.0 cal/g·°C; ΔT = 24.3 − 21.2 = 3.1 °C

q = 1000. g × 1.0 cal/g·°C × 3.1 °C = 3100 calories per gram of candy

Including the decimal point in 1000. g of water indicates that all of the zeroes are considered significant figures.

Note to teachers: The most common mistake students will make here is to use the mass of the candy bar sample, rather than the mass of the water surroundings. Remind students that the temperature change is measured for the surroundings.

(b) Energy content = $\dfrac{3100\ calories}{1.0\ g} \times \dfrac{1\ kcal}{1000\ calories} \times \dfrac{1\ Calorie}{1\ kcal} = 3.1\ Cal/g$
    (fuel value)

(c) Total calorie content = $\dfrac{3.1\ Calories}{g} \times \dfrac{454\ g}{pound} \times \dfrac{1\ pound}{16\ oz} \times 2.5\ oz = 220\ Calories$

2. Consult the nutritional labels on two of your favorite snack foods: Report their total calorie content (total Calories) and calculate their fuel value (Calories per gram).

The following nutritional information was obtained for representative snack foods:

Snack Food	Calorie Content (Calories per Serving)	Serving Size (Grams)	Fuel Value (Calories per Gram)
Fudge Cookies	150 Calories	29 g	5.2 Cal/g
Granola Bar	110 Calories	28 g	3.9 Cal/g
Cheese Crackers	190 Calories	39 g	4.9 Cal/g
Krispie Bar	90 Calories	22 g	4.1 Cal/g

Sample Data

Student data will vary.

Data Table. *The Calorimetry Experiment*

Food Sample	Initial Mass (Food Sample and Holder), g	Final Mass (Food Sample and Holder), g	Initial Temperature (Water), °C	Final Temperature (Water), °C
Peanut (1)	3.17	2.88	36.6	53.3
Peanut (2)	3.25	3.05	41.5	52.1
Cheese Puff (1)	3.08	2.62	19.3	35.8
Cheese Puff (2)	2.86	2.64	32.9	41.1
Marshmallow (1)	3.48	3.27	47.6	50.9
Marshmallow (2)	3.09	2.89	50.7	54.2
Popcorn (1)	2.77	2.65	21.7	25.3
Popcorn (2)	2.67	2.58	25.1	28.0

Teacher Notes

Answers to Post-Lab Calculations and Analysis *(Student answers will vary.)*

To summarize the results, construct a Results Table and enter all of the following information in it.

1. Determine the mass of water heated in the calorimeter for each food sample.

   The mass of the water surroundings is 50.0 g in each case.

2. Calculate the change in temperature (ΔT) for each sample.

   Sample calculation for peanuts, Run 1: $\Delta T = T_{final} - T_{initial} = 53.3\,°C - 36.6\,°C = 16.7\,°C$.

   See the Sample Results Table for all other results.

3. Use the heat equation (Equation 2) to calculate the heat (q) absorbed by the water in the calorimeter for each food sample. Report the results in *calories, kilocalories,* and *nutritional Calories.*

   Sample calculation for peanuts, Run 1: $q = m \times s \times \Delta T = 50.0\,g \times 1.0\,cal/g \cdot °C \times 16.7\,°C = 835$ calories (0.835 kilocalories or 0.835 Calories). See the Sample Results Table for all other results.

4. Subtract the final mass of the food sample and holder from the initial mass to determine the mass in grams of the food sample that burned in each experiment.

   Sample calculation for peanuts, Run 1: Mass of burned food sample $= 3.17\,g - 2.88\,g = 0.29\,g$.

   See the Sample Results Table for all other results.

5. Use the results from Questions #3 and 4 to calculate the energy content (fuel value) of the food sample in units of *Calories per gram* (Cal/g).

   Sample calculation for peanuts, Run 1: Energy content $= 0.835\,Calories/0.29\,g = 2.9\,Cal/g$.

   *See the Sample Results Table for all other results. **Note to teacher:** All results have been rounded to the appropriate number of significant figures.*

6. Record your results for the energy content of foods along with those of other groups in the class on the overhead projector or the board. Be sure to record the identity of the food sample.

The following Sample Results Table summarizes representative classroom data.

Snack Food	Mass of Water	Temperature Change	Heat Absorbed	Mass of Burned Food Sample	Energy Content
Peanut (1)	50.0 g	16.7 °C	0.835 Cal	0.29 g	2.9 Cal/g
Peanut (2)	50.0 g	10.6 °C	0.530 Cal	0.20 g	2.7 Cal/g
Cheese Puff (1)	50.0 g	16.5 °C	0.825 Cal	0.46 g	1.8 Cal/g
Cheese Puff (2)	50.0 g	8.2 °C	0.41 Cal	0.22 g	1.9 Cal/g
Marshmallow (1)	50.0 g	3.3 °C	0.17 Cal	0.21 g	0.81 Cal/g
Marshmallow (2)	50.0 g	3.5 °C	0.18 Cal	0.20 g	0.90 Cal/g
Popcorn (1)	50.0 g	3.6 °C	0.18 Cal	0.12 g	1.5 Cal/g
Popcorn (2)	50.0 g	2.9 °C	0.15 Cal	0.09 g	1.7 Cal/g

7. Copy all of the results and use the class data to calculate the *average energy content* in units of Cal/g for different types of snack foods. Construct a *Class Results Table* to summarize the results.

Sample Class Results Table

Snack Food	Average Energy Content
Peanuts	2.8 Cal/g
Cheese Puffs	1.9 Cal/g
Marshmallows	0.85 Cal/g
Popcorn	1.6 Cal/g

8. Rank the snack foods in order of their average energy content, from highest to lowest. Which snack food has the highest energy content? The lowest?

The average energy content ranged from 0.85 Cal/g for marshmallows to 2.8 Cal/g for peanuts. The average energy content of the foods from highest to lowest is:

peanuts > cheese puffs > popcorn > marshmallows

Note to teacher: *The total energy content in Calories per gram for all the foods is lower than the actual energy content listed on their nutritional label information. The relative ranking of the foods, however, from highest energy to lowest energy content, is the same as that predicted based on the information on the nutritional labels.*

The percent errors in the average energy content are quite high for most foods. This makes a good error analysis topic for post-lab discussion. What would be the energy content for a food that was 100% carbohydrate or 100% fat (4 and 9 Cal/g, respectively)?

9. Based on your knowledge of the fat content of different snack foods (if necessary, consult their nutritional labels to obtain this information), make a general statement describing the relative energy content of high-fat versus low-fat snack foods.

*Peanuts and cheese curls are relatively high-fat snack foods (Calories from fat equal 76 and 63%, respectively). Popcorn and marshmallows are low-fat snacks (Calories from fat equal 10 and 0%, respectively). The average energy content of food increases as the percent fat in the food increases. **Note to teacher:** The amount of fat in popcorn varies a great deal depending on how it was prepared. Packaged snack food popcorn has a higher fat content than freshly popped corn. Hot-air popped corn should contain zero fat.*

10. Consider the major sources of error in this experiment. Do you think your results are off on the high side or the low side? Explain.

The results for all the foods appear to be significantly lower than those calculated based on their nutritional label information.

A major source of error in the experiment is heat loss through the calorimeter. The metal calorimeter tube gets very warm throughout the calorimetry experiment, which means that the calorimeter itself, and not just the water, absorbs some of the heat given off by the burning snack food. In addition, the Erlenmeyer flask itself also absorbs a lot of heat. Heat loss through the calorimeter reduces the measured temperature change for the water surroundings, which in turn decreases the calculated value of the heat absorbed by the water. The calculated energy content values in Cal/g are likely to be lower than their actual values as a result.

A second major source of error is incomplete combustion of the food samples. All of the burning food samples produced black soot on the bottom of the Erlenmeyer flask. Also, the food samples burned for only a short time, until they became black and charred and turned to charcoal. The production of carbon means that the food molecules are not being converted to carbon dioxide and water; combustion stops at the carbon stage. This means their calculated energy contents are lower than that predicted for complete combustion. In addition, since the food samples do not burn completely, the calculated calories are not representative of the entire food sample, only of the ingredients that burned fastest.

Other minor sources of error include: (1) inadequate stirring of the water surroundings (measured temperature change is not representative of the entire water volume); and (2) the cork food holder burning during the experiment (calculated heat change includes a contribution from the heat of combustion of the cork).

Teacher Notes

Heats of Reaction and Hess's Law
Small-Scale Calorimetry

Introduction

The reaction of magnesium metal with air in a Bunsen burner flame provides a dazzling demonstration of a combustion reaction. Magnesium burns with an intense flame that produces a blinding white light. This reaction was utilized in the early days of photography as the source of "flash powder" and later in flashbulbs. It is still used today in flares and fireworks. How much heat is produced when magnesium burns?

Concepts

- Heat of reaction
- Heat of formation
- Hess's Law
- Calorimetry

Background

Magnesium reacts with oxygen in air to form magnesium oxide, according to Equation 1.

$$Mg(s) \ + \ \tfrac{1}{2}O_2(g) \ \rightarrow \ MgO(s) \ + \ heat \qquad\qquad \textit{Equation 1}$$

As mentioned above, a great deal of heat and light are produced—the temperature of the flame can reach as high as 2400 °C. The amount of heat energy produced in this reaction cannot be measured directly in the high school lab. It is possible, however, to determine the amount of heat produced by an indirect method, using Hess's Law.

The heat or enthalpy change for a chemical reaction is called the heat of reaction (ΔH_{rxn}). The enthalpy change—defined as the difference in enthalpy between the products and reactants—is equal to the amount of heat transferred at constant pressure and does not depend on how the transformation occurs. This definition of enthalpy makes it possible to determine the heats of reaction for reactions that cannot be measured directly. According to Hess's Law, if the same overall reaction is achieved in a series of steps, rather than in one step, the enthalpy change for the overall reaction is equal to the sum of the enthalpy changes for each step in the reaction series. There are two basic rules for calculating the enthalpy change for a reaction using Hess's Law.

- Equations can be "multiplied" by multiplying each stoichiometric coefficient in the balanced chemical equation by the same factor. *The heat of reaction (ΔH) is proportional to the amount of reactant.* Thus, if an equation is multiplied by a factor of two to increase the number of moles of product produced, then the heat of reaction must also be multiplied by a factor of two.

- Equations can be "subtracted" by reversing the reactants and products in the balanced chemical equation. *The heat of reaction (ΔH) for the reverse reaction is equal in magnitude but opposite in sign to that of the forward reaction.*

This experiment is designed as an advanced calorimetry lab for students who have completed a more basic calorimetry experiment. See "Measuring Energy Changes" or "Discovering Instant Cold Packs" in this lab manual.

Heats of Reaction and Hess's Law

Consider the following three reactions:

$$Mg(s) + 2HCl(aq) \rightarrow MgCl_2(aq) + H_2(g) \qquad \textit{Equation A}$$

$$MgO(s) + 2HCl(aq) \rightarrow MgCl_2(aq) + H_2O(l) \qquad \textit{Equation B}$$

$$H_2(g) + \tfrac{1}{2}O_2(g) \rightarrow H_2O(l) \qquad \textit{Equation C}$$

It is possible to express the combustion of magnesium (Equation 1) as an algebraic sum of Equations A, B, and C. Applying Hess's Law, therefore, it should also be possible to determine the heat of reaction for Equation 1 by combining the heats of reaction for Equations A–C in the same algebraic manner. *Note:* Chemical equations may be combined by addition, subtraction, multiplication, and division.

Experiment Overview

The purpose of this experiment is to use Hess's Law to determine the heat of reaction for the combustion of magnesium (Equation 1). The heats of reaction for Equations A and B will be measured by calorimetry. The heats of reaction for these two reactions will then be combined algebraically with the heat of formation of water (Equation C) to calculate the heat of reaction for the combustion of magnesium.

Pre-Lab Questions

1. Review the *Background* section. Arrange Equations A–C in such a way that they add up to Equation 1.

2. Use Hess's Law to express the heat of reaction for Equation 1 as the appropriate algebraic sum of the heats of reaction for Equations A–C.

3. The heat of reaction for Equation C is equal to the standard heat of formation of water. The heat of formation of a compound is defined as the enthalpy change for the preparation of one mole of a compound from its respective elements in their standard states at 25 °C. Chemical reference sources contain tables of heats of formation for many compounds. Look up the heat of formation of water in your textbook or in a chemical reference source such as the *CRC Handbook of Chemistry and Physics*.

Materials

Hydrochloric Acid, HCl, 1 M, 60 mL	Graduated cylinder, 25- or 50-mL
Magnesium ribbon, Mg, 7-cm strip	Metric ruler, marked in mm
Magnesium oxide, MgO, 0.40 g	Scissors
Balance, centigram (0.01 g precision)	Spatula
Calorimeter, small-scale	Stirring rod
Digital thermometer or temperature sensor	Wash bottle and water
Forceps	Weighing dish

Safety Precautions

Hydrochloric acid is toxic by ingestion and inhalation and is corrosive to skin and eyes. Magnesium metal is a flammable solid. Keep away from flames. Do not handle magnesium metal with bare hands. Wear chemical splash goggles and chemical-resistant gloves and apron. Wash hands thoroughly with soap and water before leaving the lab.

Bank statements provide an analogy that many teachers find helpful in explaining Hess's Law. Consider two students who have ending bank balances of $150 each. One student may have started with $50 in the account, made three deposits of $50 each, and two withdrawals of $25 each. The other student may have started with $50 and made only one deposit of $150 and one withdrawal of $50. Their beginning and ending bank balances were both the same and did not depend on how they got there.

Teacher Notes

Procedure

Record all data for Parts A and B in the Data Table.

Part A. Reaction of Magnesium with Hydrochloric Acid

1. Obtain a 7-cm strip of magnesium ribbon and cut it into two pieces of unequal length, roughly 3- and 4-cm each. *Note:* Handle the magnesium ribbon using forceps.

2. Measure the exact length of each piece of magnesium ribbon to the nearest 0.1 cm.

3. Multiply the length of each piece of Mg ribbon by the conversion factor (g/cm) provided by your teacher to obtain the mass of each piece of Mg.

4. Mass a clean, dry calorimeter to the nearest 0.01 g.

5. Using a graduated cylinder, add 15 mL of 1 M hydrochloric acid to the calorimeter and measure the combined mass of the calorimeter and acid.

6. Using a digital thermometer or a temperature sensor, measure the initial temperature of the hydrochloric acid solution to the nearest 0.1 °C.

7. Add the first (shorter) piece of magnesium ribbon to the acid and stir the solution until the magnesium has dissolved and the temperature of the solution remains constant.

8. Record the final temperature of the solution to the nearest 0.1 °C.

9. Rinse the contents of the calorimeter down the drain with excess water.

10. Dry the calorimeter and mass it again to the nearest 0.01 g.

11. Repeat steps 5–9 using the second (larger) piece of magnesium ribbon.

Part B. Reaction of Magnesium Oxide with Hydrochloric Acid

12. Mass a clean, dry calorimeter to the nearest 0.01 g.

13. Using a graduated cylinder, add 15 mL of 1 M HCl to the calorimeter and measure the combined mass of the calorimeter and hydrochloric acid.

14. Tare a small weighing dish and add about 0.20 g of magnesium oxide. Measure the exact mass of magnesium oxide to the nearest 0.01 g.

15. Using a digital thermometer or a temperature sensor, measure the initial temperature of the hydrochloric acid solution to the nearest 0.1 °C.

16. Using a spatula, add the magnesium oxide to the acid. Stir the reaction mixture until the temperature remains constant for several five-second intervals. Record the final temperature of the solution to the nearest 0.1 °C.

17. Pour the reaction mixture down the drain with excess water. Rinse and dry the calorimeter.

18. Repeat steps 12–16 using a second sample of magnesium oxide.

19. Wash the contents of the calorimeter down the drain with excess water.

In order to obtain two significant figures in the calculations involving the mass of Mg, it is necessary to give the students a conversion factor in g/cm for the mass of Mg. Measure the exact length and exact mass of a 20-cm piece of Mg ribbon. Divide the mass of the ribbon by its length to obtain the conversion factor. Example: A 20.0 cm piece of Mg weighed 0.19 g. The conversion factor is 0.0095 g/cm. Notice two significant figures are allowed in the recorded mass.

Heats of Reaction and Hess's Law

Name: _____

Class/Lab Period: _____

Heats of Reaction and Hess's Law

Data Table

	Reaction A (Mg + HCl)		Reaction B (MgO + HCl)	
	Trial 1	**Trial 2**	**Trial 1**	**Trial 2**
Mass of Calorimeter (g)				
Mass of Calorimeter + HCl Solution (g)				
Mass of Mg (Reaction A) or MgO (Reaction B) (g)				
Initial Temperature (°C)				
Final Temperature (°C)				

Post-Lab Calculations and Analysis *(Show all work on a separate sheet of paper.)*

Construct a Results Table to summarize the results of all calculations. For each reaction and trial, calculate the:

1. Mass of hydrochloric acid solution.

2. Total mass of the reactants.

3. Change in temperature, $\Delta T = T_{final} - T_{initial}$.

4. Heat (q) absorbed by the solution in the calorimeter. *Note:* $q = m \times s \times \Delta T$, where s is the specific heat of the solution in J/g·°C. Use the total mass of reactants for the mass (m) and assume the specific heat is the same as that of water, namely, 4.18 J/g·°C.

5. Number of moles of magnesium and magnesium oxide in Reactions A and B, respectively.

6. Enthalpy change for each reaction in units of kilojoules per mole (kJ/mole).

7. Average enthalpy change (heat of reaction, ΔH_{rxn}) for Reactions A and B. *Note:* The enthalpy change is positive for an endothermic reaction, negative for an exothermic reaction.

8. Use Hess's Law to calculate the heat of reaction for Equation 1. *Hint:* See your answer to PreLab Question #2.

9. The heat of reaction for Equation 1 is equal to the heat of formation of solid magnesium oxide.

   (a) Look up the heat of formation of magnesium oxide in your textbook or a chemical reference source.

   (b) Calculate the percent error in your experimental determination of the heat of reaction for Equation 1.

Teacher Notes

Teacher's Notes

Heats of Reaction and Hess's Law

Master Materials List *(for a class of 30 students working in pairs)*

Hydrochloric acid, HCl, 1 M, 900 mL	Graduated cylinders, 25- or 50-mL, 15
Magnesium ribbon, Mg, 7-cm strips, 15	Metric rulers, 15
Magnesium oxide, MgO, 6 g	Scissors, 3
Balances, centigram (0.01 g precision), 3	Spatulas, 15
Calorimeters, small-scale, 15*	Stirring rods, 15
Digital thermometers or temperature sensors, 15	Wash bottles and water, 15
Forceps, 15	Weighing dishes, 30

*Small-scale calorimeters are available from Flinn Scientific, Catalog No. AP5928; see the *Lab Hints* Section.

Preparation of Solutions *(for a class of 30 students working in pairs)*

Hydrochloric Acid, 1 M. Place about 500 mL of distilled or deionized water in a flask and add 83 mL of 12 M hydrochloric acid. Stir to mix and then dilute to 1 L. *Note:* Always add acid to water.

Safety Precautions

Hydrochloric acid is toxic by ingestion and inhalation and is corrosive to skin and eyes. Magnesium metal is a flammable solid. Keep away from flames. Do not handle magnesium metal with bare hands. Wear chemical splash goggles and chemical-resistant gloves and apron. Wash hands thoroughly with soap and water before leaving the lab. Consult current Material Safety Data Sheets for additional safety, handling, and disposal information.

Disposal

Please consult your current *Flinn Scientific Catalog/Reference Manual* for general guidelines and specific procedures governing the disposal of laboratory waste. The final solutions may be disposed of down the drain with excess water according to Flinn Suggested Disposal Method #26b.

Lab Hints

For best results, use fresh magnesium ribbon. If the metal ribbon appears to be oxidized, rub it with some steel wool to remove the oxide coating. Measure the mass of magnesium ribbon after it has been cleaned, if necessary.

- The experimental work for this lab can reasonably be completed in one 50-minute period. Each trial should take no more than 10 minutes. If time is a concern, consider performing the experiment as a cooperative class activity, in which each group performs two trials of either Reaction A or Reaction B. Students calculate the heats of reaction for their two trials and record their results, along with those of the rest of the class, on the board or overhead projector. The class results for both Reactions A and B are averaged and the average heats of reaction are used to calculate the heat of reaction for Equation 1.

- The small-scale calorimeters recommended in the *Materials* section are available from Flinn Scientific, Inc. (Catalog No. AP5928). These calorimeters are manufactured from dense, 2″-thick polystyrene and lined with a specially formulated coating that permeates the pores of the foam. The lining maximizes heat efficiency and limits water absorption, which means that the calorimeter constant is small and does not change. The maximum temperature recommended for these small-scale calorimeters is 50 °C. Small-scale calorimeters work well with the microscale quantities used in this experiment.

- The thermometer may be used as the stirrer with small-scale calorimeters because the soft foam is not likely to damage any type of thermometer. Advise students, however, not to punch holes or indentations into the bottom or sides of the calorimeter. These indentations may trap liquid and thus interfere with both mixing the solution and drying the calorimeter for multiple trials.

- Because of the small-scale nature of this experiment, the maximum time required for any given trial, including measuring and weighing the reagents, is less than 10 minutes. This factor allows students to carry out several trials in a single class period. The maximum temperatures in this experiment are usually reached within one minute after mixing and are stable for at least 20–30 seconds. The temperatures then begin to decrease at a rate of about 0.1 °C every 20 seconds.

- The best thermometers for small-scale calorimetry are digital electronic thermometers (Flinn Scientific Catalog No. AP8716) or temperature sensors connected to a computer- or calculator-based interface system such as LabPro or CBL. Digital thermometers are reasonably inexpensive, update every second, and are precise to the nearest 0.1 °C. Temperature measurements may be a significant source of error in calorimetry experiments. The *Supplementary Information* Section contains instructions for adapting this experiment to the use of technology for computer-interfaced data collection and analysis.

- For greater accuracy in the results, add a correction term to each enthalpy calculation to account for the heat loss to the calorimeter. See the *Supplementary Information* section.

- The procedure may be scaled up if conventional Styrofoam™ cup calorimeters are used. Use 100 mL of hydrochloric acid per trial and increase the amount of reagents to 0.5 g of magnesium ribbon and 1.0 g of magnesium oxide in Parts A and B, respectively.

Teaching Tips

- This experiment is designed as an advanced lab activity for students who have completed a basic, introductory calorimetry experiment, such as "Measuring Energy Changes," "Discovering Instant Cold Packs," or "Measuring Calories" in this Flinn ChemTopic™ Labs manual. For a review of the basics of calorimetry, see the *Supplementary Information* section.

- Consider leading into this experiment with a visual demonstration of the combustion of magnesium ribbon in a Bunsen burner flame. This simple demonstration will arouse interest and provides a counterbalance against the abstract nature of Hess's Law. When performing the demonstration for the students, instruct students not to look directly or

Teacher Notes

stare at the bright white flame. The bright light emitted by the burning magnesium is UV light and can damage the eyes. Students should look at the flame "out of the corners of their eyes," using their peripheral vision.

- Using the total mass of the reaction mixture (hydrochloric acid solution plus magnesium or magnesium oxide) in the heat equation calculations ($q = m \times s \times \Delta T$) may be confusing to some students. Students may argue that they are measuring the temperature increase in the surroundings, not the system, and thus they should not include the mass of the reactants and products. Using the combined mass of the reaction mixture is traditional in these types of calorimetry exercises and may compensate for the fact that the specific heat of the solution is assumed to be equal to 4.18 J/g·°C, the same as that of water.

- What makes Hess's Law a law? This may be a good time to review with students the definition of a natural law. A law is not engraved in stone in nature—it is the expresssion of the results of many experiments repeated for many different systems. The "law" is a generalization that has been widely tested and has been found to be true for every reaction that has been tested. Hess's Law is also known as the Law of Additivity of Reaction Heats.

Answers to Pre-Lab Questions *(Student answers will vary.)*

1. Review the *Background* section. Arrange Equations A–C in such a way that they add up to Equation 1.

$$Mg(s) + 2HCl(aq) \rightarrow MgCl_2(aq) + H_2(g) \qquad \textit{Equation A}$$

$$MgCl_2(aq) + H_2O(l) \rightarrow MgO(s) + 2HCl(aq) \qquad \textit{– Equation B}$$

$$\underline{H_2(g) + \tfrac{1}{2}O_2(g) \rightarrow H_2O(l)} \qquad \textit{+ Equation C}$$

$$Mg(s) + \tfrac{1}{2}O_2(g) \rightarrow MgO(s) \qquad \textit{Equation 1}$$

2. Use Hess's Law to express the heat of reaction for Equation 1 as the appropriate algebraic sum of the heats of reaction for Equations A–C.

$$\Delta H_A - \Delta H_B + \Delta H_C = \Delta H_1$$

3. The heat of reaction for Equation C is equal to the standard heat of formation of water. The heat of formation of a compound is defined as the enthalpy change for the preparation of one mole of a compound from its respective elements in their standard states at 25 °C. Chemical reference sources contain tables of heats of formation for many compounds. Look up the heat of formation of water in your textbook or in a chemical reference source such as the *CRC Handbook of Chemistry and Physics*.

   The standard heat of formation of liquid water at 25 °C is equal to –285.8 kJ/mole at 25 °C. (CRC Handbook of Chemistry and Physics, *82nd Edition, 2001*).

It is very important that students complete the Pre-Lab Questions before doing the experiment. We recommend that the teacher check the students' work before allowing them to do the lab. Students who do not understand the Hess's Law calculations will not really "get" the experiment.

Sample Data

Student data will vary.

Data Table

	Reaction A (Mg + HCl)		Reaction B (MgO + HCl)	
	Trial 1	Trial 2	Trial 1	Trial 2
Mass of Calorimeter (g)	2.38	2.43	2.21	2.21
Mass of Calorimeter + HCl Solution (g)	16.97	16.84	17.23	17.15
Mass of Mg (Reaction A) or MgO (Reaction B) (g)	0.028	0.038	0.20	0.18
Initial Temperature (°C)	21.2	21.2	21.1	21.1
Final Temperature (°C)	30.0	32.4	31.0	30.6

Answers to Post-Lab Calculations and Analysis *(Student answers will vary.)*

Construct a Results Table to summarize the results of all calculations. For each reaction and trial, calculate the:

1. Mass of hydrochloric acid solution.

   Sample calculation for Reaction A, Trial 1: 16.97 g – 2.38 g = 14.59 g. See Sample Results Table on page 60 for results of all other calculations.

2. Total mass of the reactants.

   Sample calculation for Reaction A, Trial 1: 14.59 g + 0.03 g = 14.62 g. See Sample Results Table on page 60 for results of all other calculations.

3. Change in temperature, $\Delta T = T_{final} - T_{initial}$.

   Sample calculation for Reaction A, Trial 1: 30.0 °C – 21.2 °C = 8.8 °C. See Sample Results Table on page 60 for results of all other calculations.

4. Heat (q) absorbed by the solution in the calorimeter. *Note:* $q = m \times s \times \Delta T$, where s is the specific heat of the solution in J/g·°C. Use the total mass of reactants for the mass (m) and assume the specific heat is the same as that of water, namely, 4.18 J/g·°C.

   Sample calculation for Reaction A, Trial 1: q = 14.62 g × 4.18 J/g·°C × 8.8 °C = 540 J. See Sample Results Table on page 60 for results of all other calculations.

5. Number of moles of magnesium and magnesium oxide in Reactions A and B, respectively.

   Sample calculation for Reaction A, Trial 1: 0.028 g Mg × (1 mole/24.3 g) = 1.2×10^{-3} moles Mg. See Sample Results Table on page 60 for results of all other calculations.

6. Enthalpy change for each reaction in units of kilojoules per mole (kJ/mole).

   Sample calculation for Reaction A, Trial 1: $-540 \text{ J}/(1.2 \times 10^{-3} \text{ moles}) \times 1 \text{ kJ}/1000 \text{ J}$ = -450 kJ/mole. Notice the negative sign for the enthalpy change for the reaction. Reactions A and B are both exothermic reactions. The heat absorbed by the solution in the calorimeter is equal in magnitude but opposite in sign to the heat released by the reaction. See Sample Results Table on page 60 for results of all other calculations.

7. Average enthalpy change (heat of reaction, ΔH_{rxn}) for Reactions A and B. *Note:* The enthalpy change is positive for an endothermic reaction, negative for an exothermic reaction.

   Sample calculation for Reaction A: $[-450 + (-420)]/2 = -440 \text{ kJ/mole}$. See Sample Results Table on page 60 for results of all other calculations.

8. Use Hess's Law to calculate the heat of reaction for Equation 1. *Hint:* See your answer to PreLab Question #2.

   $$\Delta H_1 = \Delta H_A - \Delta H_B + \Delta H_C$$

   $$\Delta H_1 = [-440 - (-130) + (-286)] \text{ kJ/mole} = -600 \text{ kJ/mole} \text{ (rounded to two significant figures)}$$

9. The heat of reaction for Equation 1 is equal to the heat of formation of solid magnesium oxide.

   (a) Look up the heat of formation of magnesium oxide in your textbook or a chemical reference source.

   (b) Calculate the percent error in your experimental determination of the heat of reaction for Equation 1.

   (a) The standard heat of formation of solid magnesium oxide at 25 °C is equal to -601.6 kJ/mole. (CRC Handbook of Chemistry and Physics, *82nd Edition, 2001*).

   (b) $$\text{percent error} = \frac{|experimental - theoretical|}{theoretical} \times 100\% = \frac{|-600 - (-602)|}{602} = 0.3\%$$

The remarkable accuracy of these tested results is due to the use of the special, small-scale calorimeters. Expect much greater percent errors in this experiment if conventional coffee-cup calorimeters are used.

Sample Results Table

	Reaction A (Mg + HCl)		Reaction B (MgO + HCl)	
	Trial 1	Trial 2	Trial 1	Trial 2
Mass of Hydrochloric Acid	14.59 g	14.41 g	15.02 g	14.94 g
Total Mass of Reactants	14.62 g	14.45 g	15.22 g	15.12
Temperature Change	8.8 °C	11.2 °C	9.9 °C	9.5 °C
Heat Absorbed by Solution	540 J	676 J	630 J	600 J
Moles of Mg or MgO	.0012 moles	.0016 moles	.0050 moles	.0045 moles
Enthalpy Change per Mole of Mg or MgO§	–450 kJ/mole	–420 kJ/mole	–126 kJ/mole	–133 kJ/mole
Average Enthalpy Change	–440 kJ/mole		–130 kJ/mole	

§The calculated enthalpy changes are all negative values. These are both exothermic reactions—the heat released by the system resulted in a temperature increase in the surroundings. The sign of the enthalpy change is a common source of student error.

Supplementary Information

Calorimetry Basics

Calorimetry experiments are carried out by measuring the temperature change in water that is in contact with or surrounds the reactants and products. In a typical calorimetry experiment, the reaction of a known mass of reactant is carried out either directly in or surrounded by a known quantity of water and the temperature increase or decrease in the surrounding water is measured. The temperature change (ΔT) produced in the water is related to the amount of heat energy (q) absorbed or released by the reaction system according to the following equation:

$$q = m \times s \times \Delta T$$

where m is the mass of the aqueous solution, s is the specific heat of water, and ΔT is the observed temperature change. The specific heat of water is defined as the amount of heat required to increase the temperature of one gram of water by 1 °C and is equal to 4.18 cal/g·°C.

Measuring the Calorimeter Constant of the Calorimeter

When equal volumes of hot and cold water are combined, the new temperature should be the average of the two initial temperatures if there is no heat loss to the calorimeter. In actual practice, the new temperature will be slightly less than the average due to heat loss to the calorimeter. Use the following procedure to determine the calorimeter constant in J/°C.

1. Label two calorimeters "cold water" and "warm water." Mass each calorimeter to the nearest 0.01 g. Add 8 mL of cold tap water to the cold water calorimeter and find its mass to the nearest 0.01 g. Add 8 mL of warm tap water to the warm water calorimeter and find its mass to the nearest 0.01 g.

2. Stir the water in each calorimeter and record the initial temperature in each calorimeter to the nearest 0.1 °C (the initial temperature should be stable for at least 20 sec).

3. Pour the cold water into the warm water calorimeter and record the resulting final temperature of the mixture to the nearest 0.1 °C.

4. Rinse and dry the calorimeter and repeat at least once.

5. Calculate the mass of water in each calorimeter.

6. Subtract the initial temperature of the cold and warm water from the final temperature to determine the temperature change for both the cold and warm water (ΔT_{cold} and ΔT_{warm}).

7. Calculate the energy gained by the cold water: $q_{cold} = m_{cold} \times s \times \Delta T_{cold}$.

8. Calculate the energy lost by the warm water: $q_{warm} = m_{warm} \times s \times \Delta T_{warm}$.

9. Determine the absolute difference between the energy lost by the warm water and the energy gained by the cold water. Calculate the calorimeter constant in J/°C using the following equation:

$$\text{calorimeter constant} = \frac{|q_{warm} - q_{cold}|}{\Delta T_{warm}}$$

10. For each trial in Parts A and B, multiply the observed temperature change ΔT by the calorimeter constant to determine the heat absorbed by the calorimeter.

11. In calculating the heat released by the exothermic reactions in Parts A and B, add a term to correct for the heat absorbed by the calorimeter:

heat released by reaction = − (heat absorbed by solution + heat absorbed by calorimeter)

Alternative Procedure
(Computer-Interfaced Data Collection and Analysis)

The following instructions are provided for adapting the experiment to the use of technology (computer-interfaced data collection and analysis).

1. Connect the interface (LabPro, CBL system, etc.) to a computer or calculator.

2. Plug a temperature probe into the interface.

3. Open and set up a graph in your data collection software so that the y-axis reads temperature in degrees Celsius. Set the minimum and maximum temperature values at 20 degrees and 50 degrees, respectively.

4. The x-axis should be set for time in minutes. Set the minimum and maximum time values at 0 seconds and 240 seconds, respectively.

5. The time interval should be set so a temperature reading is taken every 10 seconds.

6. Obtain 15 mL of hydrochloric acid in a graduated cylinder and carefully transfer the acid to the calorimeter.

7. Place the temperature probe in the acid solution. Allow the probe to equilibrate at the initial temperature for 1 minute, then press Start to begin collecting temperature data.

8. Stir the solution with the temperature sensor and add the pre-weighed amount of magnesium ribbon or magnesium oxide in Parts A and B, respectively.

9. Continue stirring the solution and collecting temperature data for 4 minutes (240 seconds).

10. The system will automatically record data for the allotted time (240 seconds), then stop.

11. Print the computer-generated data table and graph, if possible, and use the data to complete the Data Table and the Post-Lab Calculations.

Colorful Heat
Heat versus Temperature Demonstration

Introduction

Students use the terms heat and temperature interchangeably in their daily lives. Scientifically, however, these two terms represent different concepts. This demonstration provides a colorful illustration of the relationship between heat and temperature—a key concept in thermochemistry.

Concepts

- Temperature
- Heat

Materials

Beakers, 250-mL, 5
Food coloring, red

Stirring rod
Water

Safety Precautions

Although the materials in this demonstration are considered nonhazardous, observe all normal laboratory safety guidelines.

Procedure

1. Place five 250-mL beakers on the demonstration table and add the following amounts of water to each:

Beaker	1	2	3	4	5
Water (mL)	10 mL	50 mL	100 mL	150 mL	200 mL

2. Carefully add one drop of red food coloring to each beaker and stir. *(One drop of food coloring represents a standard amount of heat added to each beaker.)*

3. Place a piece of white paper behind the beakers and view the red solution from the front, not the top. *(The intensity of the red color represents the temperature of the liquid.)*

4. Which beaker has the most "heat" (food coloring) in it? *(All five beakers contain the same amount of heat.)*

5. Which beaker has the highest "temperature" (color intensity)? *(Beaker #1 has the highest temperature.)*

6. Ask students to propose a relationship between the amount of heat added to a substance and the resulting temperature increase. *(This demonstration looks at the relationship between the amount of heat absorbed and the resulting temperature change as a function of the amount of the substance (volume of water). The temperature increase depends on the volume (mass) of water in the beaker. Another factor that influences the resulting temperature change is the ability of a substance to absorb heat, defined as its specific heat. This property is the same in each beaker.)*

7. Based on the proposed relationship between heat and temperature, how much "heat" must be added to beaker #5 (200 mL of water) to give it the same "temperature" as beaker

Demonstrations

#2 (50 mL of water)? *(Students should deduce that they would have to add four times as much heat because the volume is four times greater.)* Do the experiment!

8. How much "heat" must be added to beaker #3 (100 mL of water) to give it the same "temperature" as beaker #1 (10 mL of water)? Do the experiment!

Disposal

The solutions may be rinsed down the drain with excess water. Please consult your current *Flinn Scientific Catalog/Reference Manual* for specific disposal procedures.

Tips

- The concepts developed in this demonstration are complementary to those in the "Specific Heat" demonstration that follows in this *Flinn ChemTopic™ Labs* book. "Specific Heat" illustrates the relationship between heat and specific heat.

- Large volumes of water have the ability to store large amounts of heat at moderate temperatures. Water is used in solar heated homes to store energy from the sun. Energy absorbed during the day from the sun's rays is used to heat water in a reservoir. At night, the energy released from the water reservoir is used to heat the air that circulates throughout the home.

Discussion

Students often mistakenly assume that temperature and heat are the same thing. Part of the confusion results from common usage of the words hot and cold. A drop of water from a boiling tea kettle and a cup of hot tea, after all, both may be "hot"—they are at the same temperature. Spill a drop of hot water on your hand and it will burn a little. Spill a cup of tea on your hand, however, and it's off to the emergency room. The difference is not the temperature of the water, but its heat content.

Heat is a form of energy. Heat is defined as the amount of energy transferred from a substance at a higher temperature to a substance at a lower temperature. The best working definition of temperature is that objects in thermal equilibrium must be at the same temperature. Temperature is a quantitative measure of heat intensity on some defined scale (Celsius, Fahrenheit, Kelvin). On a molecular level—according to the kinetic theory—the absolute temperature of a substance in kelvins (degrees Kelvin) is proportional to the average kinetic energy of molecules.

The amount of heat (q) transferred to an object or substance depends on three factors: the amount of the substance (its mass, m), the ability of the substance to absorb heat (its specific heat, s), and the resulting temperature change (ΔT), according to Equation 1.

$$q = m \times s \times \Delta T \qquad \textit{Equation 1}$$

The SI unit for energy (and heat) is the joule (abbreviated J). Specific heat—defined as the amount of heat needed to raise the temperature of one gram of a substance by 1 °C—is a characteristic physical property of a substance. The specific heat of water is equal to 4.18 Joules per gram per degree Celsius (4.18 J/g·°C).

Equation 1 may be introduced within the context of this demonstration. Steps 7 and 8 reveal that the amount of heat that must be added to water to produce a given temperature change is proportional to the mass of water used.

Specific Heat
Chemistry Demonstration

Introduction

Three different metals of equal mass are heated to the same temperature in a boiling water bath. The metals are then added to three insulated foam cups, each containing the same amount of water. The resulting temperature changes are different for the three metals. What causes the difference?

Concepts

- Heat
- Specific heat
- Heat capacity
- Calorimetry

Materials

Metal shot, Al, Pb, and Zn, 40–60 g each*	Hot plate or Bunsen burner setup
Balance, centigram (0.01 g precision)	Insulated foam (Styrofoam®) cups, 6
Beaker, 600-mL	Stirring rods, 3
Beakers, 400-mL, 3	Test tubes, large (25 × 150 mm), 3
Boiling stones	Test tube holder or tongs
Digital thermometers, 3	Water
Graduated cylinder, 100-mL	

*The "Specific Heat Set" (AP9220) is ideal for this activity. See the *Tips* Section.

Safety Precautions

Handle hot metal samples with care to avoid burns. Wear chemical splash goggles whenever working with chemicals, heat, or glassware in the lab.

Procedure

Recruit student volunteers to measure and record data.

1. Prepare a boiling water bath: Fill a large, 600-mL beaker half-full with water, add a couple of boiling stones, and heat the water to boiling using a hot plate or Bunsen burner setup.

2. Prepare equal-mass (±0.2 g) samples, 40–60 g each, of the three metals. Record the mass and identity of each metal.

3. Label three large test tubes and add the metal samples to them. Place the test tubes in the boiling water bath for 10 minutes to heat them to 100 °C.

4. Prepare three calorimeters: Nest two Styrofoam cups together and set them in a 400-mL beaker for extra stability. Label the calorimeters. Using a graduated cylinder, add 50 mL of water to each calorimeter. Measure the initial temperature of water.

5. Using a test tube holder, lift each test tube from the boiling water bath and quickly, yet carefully, pour the metal sample into the appropriately labeled calorimeter.

Because the density of aluminum is much lower than that of lead and zinc, an equal mass of Al occupies a much larger volume than Pb or Zn. Choose a mass such that the aluminum shot in the test tube will be submerged in the boiling water bath.

6. Stir the water in each calorimeter with a stirring rod. Measure the highest (final) temperature the water reaches.

7. Compare the final temperature of the water in the three calorimeters. Are they the same? Why or why not? What variables have been controlled in each calorimeter experiment? What variables are different? *[The following variables were controlled in this experiment: the volume (mass) and initial temperature of the water, the initial temperature of the metals, the mass of the metals, the type of reaction vessel. The variable that is different is the identity and nature of the metal. In order for the final temperature of the water to be different in the three calorimeters, different amounts of heat must have been added. Students should conclude that different substances—in this case metals—store different amounts of heat at the same temperature. The ability of a substance to store heat is one way of defining the heat capacity or specific heat of a substance.]*

8. *(Optional)* Use the heat equation: $q = m \times s \times \Delta T$, to calculate the amount of heat gained by the water in each calorimeter. *Note:* Use the mass and specific heat of water (4.18 J/g·C) in this calculation.

9. *(Optional)* According to the law of conservation of energy, the energy gained by the water in each calorimeter is equal to the amount of heat lost by the metal sample as it cooled. Calculate the amount of heat lost by each metal sample.

10. *(Optional)* Use the heat equation to calculate the specific heat of each metal. *Note:* Use the mass of the metal and the temperature change of the metal in this calculation. Assume that the initial temperature of the metal is 100 °C and the final temperature of the metal is the same as the final temperature of the water in the calorimeter.

11. *(Optional)* Discuss the findings. If the metal samples were not identified at the beginning of the activity, hand out a reference table listing the specific heats of different metals and ask students to identify the metals.

Disposal

The water may be rinsed down the drain. The metal samples should be collected, dried, and stored for repeat use. Please consult your current *Flinn Scientific Catalog/Reference Manual* for storage information and disposal procedures.

Tips

- Choose metals that will give a range of temperature changes. Aluminum, zinc, and lead are good choices: their specific heat values (in units of J/g·°C) are high (0.899), medium (0.385), and low (0.129), respectively. Copper and tin shot are also suitable.

- The "Specific Heat Set" available from Flinn Scientific, Inc. (Catalog No. AP9220) contains five different metal cylinders (aluminum, copper, lead, tin, and zinc) of equal mass (58.2 ±0.1 g). The pre-massed metal cylinders are convenient, easy-to-use, and give a wide range of temperature changes.

- Prepare the boiling water bath and calorimeters and pre-measure the metal samples before class begins. It is important that students see, however, that the metals have the same mass and that the amount of water in each calorimeter is the same.

- This demonstration may be used on a qualitative level (Steps 1–7) to illustrate the definition of specific heat, or it may be used to introduce a quantitative in-class activity (Steps 8–11). If specific heat and heat equation calculations are included in your curriculum, this demonstration will make them seem more real, less contrived, than sample problems from a book.

- Discuss the specific heats of different metals in terms of their use in cookware (pots and pans). Is a low or high specific heat desired in cookware? What other properties of the metals are important in this application?

Discussion

Transfer of heat or heat flow always occurs in one direction—from a region of higher temperature to a region of lower temperature—until some final equilibrium temperature is reached. In this demonstration, heat is transferred from a hot metal sample to colder water sample. Each metal causes the temperature of the water to increase to a different extent. This means that each metal must have a different ability to absorb heat energy and release it to the water.

The ability of a substance to contain or absorb heat energy is called its heat capacity. Heat capacity is an extensive property—it depends on the amount or mass of the sample. Specific heat is a measure of the heat capacity of a substance. Specific heat is defined as the amount of heat required to increase the temperature of one gram of a substance by one degree Celsius. It is an intensive property, a characteristic physical property of a substance. Specific heat is usually represented by the symbol s and is given in SI units of J/g·°C. Table 1 lists the specific heats of six metals that may be used in this demonstration. Notice a general trend— the larger the atomic mass of a metal, the lower its specific heat. Copper and zinc have very similar atomic masses, and identical specific heat values.

Table 1.

Metal	Aluminum	Copper	Lead	Silver	Tin	Zinc
Specific Heat (J/g·°C)	0.899	0.385	0.129	0.234	0.222	0.385

Water is a very unusual substance in many ways. The specific heat of water is very high, 4.18 J/g·°C. The high specific heat value of water means that water is able to absorb and store large amounts of heat.

The amount of heat (q) transferred to an object or substance depends on three factors: the amount of the substance (its mass, m), the ability of the substance to absorb heat (its specific heat, s), and the resulting temperature change (ΔT), according to Equation 1. The SI unit of heat (and energy) is the joule (abbreviated J).

$$q = m \times s \times \Delta T \qquad \qquad \textit{Equation 1}$$

Equation 1 may be used to calculate the specific heats of the metals used in this demonstration (see Steps 8–11). The heat gained by the water in the calorimeter should be equal in magnitude but opposite in sign to the heat lost by the metal sample in each case:

$$q(\text{gained by water}) = -q(\text{lost by metal}) \qquad \qquad \textit{Equation 2}$$

The amount of heat gained by the water is calculated using the mass of water, its specific heat, and the difference between the final and initial temperature of the water in the cup:

$$q\text{(gained by water)} = \text{mass (water)} \times s\text{ (water)} \times \Delta T\text{ (water)} \qquad \textit{Equation 3}$$

The amount of heat lost by the metal is calculated using the mass of the metal, its specific heat, and the difference between its final temperature (assumed to be the same as the final temperature of the water in the cup) and its initial temperature (assumed to be equal to the temperature of the boiling water bath in Step 3):

$$q\text{(lost by metal)} = \text{mass (metal)} \times s\text{ (metal)} \times \Delta T\text{ (metal)} \qquad \textit{Equation 4}$$

Combining Equations 2–4 makes it possible to calculate the specific heats of the metals using the data obtained in this demonstration. Sample data and results are reported in Table 2 for zinc and aluminum.

Table 2.

	Zinc	Aluminum
Mass of Metal	46.19 g	46.31 g
Mass of Water	50.0 g	50.0 g
Initial Temperature (Calorimeter)	20.0 °C	20.0 °C
Final Temperature (Calorimeter)	26.2 °C	32.9 °C
Initial Temperature (Metal)	100 °C	100 °C
ΔT (Water)	6.2 °C	12.9 °C
ΔT (Metal)	73.8 °C	67.1 °C
Specific Heat (Metal), calculated	0.38 J/g·°C	0.868 J/g·°C
Specific Heat (Metal), literature	0.385	0.899
Percent Error	1%	4%

The Cool Reaction
An Endothermic Demonstration

Introduction

Many reactions in our daily lives are exothermic, that is, they produce heat. In fact, when most people think of chemical reactions, they think of reactions that give off lots of heat (and often light and sound as well). Reactions that consume heat—called endothermic reactions—can be just as exciting. One of the most striking examples of an endothermic reaction occurs when the solids barium hydroxide and ammonium thiocyanate are mixed together in an Erlenmeyer flask.

Concepts

- Endothermic reaction
- Heat of reaction
- Enthalpy
- Entropy

Materials

Ammonium thiocyanate, NH_4SCN, 10 g

Barium hydroxide octahydrate, $Ba(OH)_2 \cdot 8H_2O$, 20 g

Erlenmeyer flask with stopper, 125-mL

Stirring rod

Gloves, Zetex,™ for low temperatures

Thermometer, graduated to at least –30 °C

Wash bottle and water

Wood block, small, such as a 6–8 inch long section of a 2×4

Safety Precautions

Barium salts are toxic by ingestion. Ammonium thiocyanate is also toxic by ingestion. Use caution when handling the cold beaker or flask—wear protective gloves. The temperatures involved are cold enough to freeze skin. Ammonia vapor is produced in the reaction; it is very irritating to eyes and the respiratory tract. Do not allow students to inhale this gas. Wear chemical splash goggles and chemical-resistant gloves and apron. Please review current Material Safety Data Sheets for additional safety, handling, and disposal information.

Procedure

1. Transfer pre-weighed portions of barium hydroxide (20 g) and ammonium thiocyanate (10 g) to the Erlenmeyer flask and mix the two solids with a glass or plastic stirring rod.

2. Within about 30 seconds, the odor of ammonia will be detected from the flask and a noticeable amount of liquid will form. The flask becomes cold to the touch and, in a humid environment, frost collects on the outside. *Caution:* If students are allowed to check for the presence of ammonia vapor, warn them not to inhale large amounts of the pungent gas.

"The Cool Reaction" is available as a Chemical Demonstration Kit from Flinn Scientific (Catalog No. AP8896).

3. Use a thermometer to measure the temperature of the liquid. The temperature drops from room temperature to about –25 °C to –30 °C within 1–2 minutes after mixing and remains below –20 °C for several minutes.

4 To provide a dramatic demonstration of the observed temperature decrease, wet a wooden block with a few drops of water and place the stoppered flask on it. The flask will quickly freeze to the wood block. Ask a student volunteer to put on protective gloves and pick up the flask—both the flask and block will be lifted off the table together!

Disposal

The products of this reaction may be disposed of according to Flinn Suggested Disposal Method #27h. Please consult your current *Flinn Scientific Catalog/Reference Manual* for general guidelines and specific procedures governing the disposal of laboratory waste.

Tips

- A 50-mL beaker and Parafilm® M may be used instead of the Erlenmeyer flask and stopper.

- Similar reactions occur with other ammonium salts, including ammonium chloride and ammonium nitrate. Substitute 7 g and 10 g of ammonium chloride and ammonium nitrate, respectively, for the 10 g of ammonium thiocyanate called for in the *Materials* Section.

- Thermodynamic data is available for the reaction between barium hydroxide and ammonium chloride or ammonium nitrate. As an extension of this activity, energetic students (no pun intended) may be motivated to write equations for the reactions and calculate the theoretical heats of reaction. Their results can then be tested using calorimetry.

Discussion

This demonstration is a great way to introduce the concept of heat transfer in chemical reactions. In this reaction, heat is absorbed from the surroundings as the reaction proceeds. The surroundings lose so much heat that water freezes. The following balanced equation summarizes the reaction between barium hydroxide and ammonium thiocyanate:

$$Ba(OH)_2 \cdot 8H_2O(s) + 2NH_4SCN(s) + heat \rightarrow Ba(SCN)_2(aq) + 2NH_3(aq) + 10H_2O(l)$$

This demonstration may also be used in an expanded discussion of thermodynamics to illustrate the relationship between the enthalpy and entropy changes for a reaction and reaction spontaneity. Many students wrongly assume that for a reaction to be spontaneous, it must be exothermic. In fact, an exothermic reaction is neither necessary nor sufficient for a reaction to be spontaneous. In the case of the reaction of barium hydroxide with ammonium thiocyanate, the large, positive enthalpy change for the endothermic reaction is more than offset by a large increase in entropy. The entropy increase in going from reactants to products is related to an increase in the number of particles and the change in state from solids to liquids. There are more product particles than reactant particles, and the products consist of a liquid and a dissolved gas (ammonia) rather than solids.

Whoosh Bottle
Chemical Demonstration

Introduction

Combustion reactions represent the most important application of thermochemistry in our daily lives. We rely on the energy produced by the combustion of fossil fuels to heat our homes and power our vehicles. How much energy is released when fuels burn? In this classic "whoosh bottle" demonstration, students observe the dramatic "whoosh" of light, sound, and heat in the combustion of isopropyl alcohol and discover the products and heat of combustion.

Concepts

- Exothermic reaction
- Combustion reactions
- Heat of combustion
- Heat of formation

Materials

Isopropyl alcohol, $(CH_3)_2CHOH$, 25 mL	Matches and wood splints
Funnel, small	Meter stick
Fire blanket and fire extinguisher	Safety shield
Graduated cylinder, 25-mL	"Whoosh bottle," plastic jug, 5-gallon

Safety Precautions

Please read all safety precautions before proceeding with this demonstration. Isopropyl alcohol is a flammable liquid and a fire hazard. It is slightly toxic by ingestion and inhalation. Use in a well-ventilated room. Recap the alcohol bottle and remove it from the demonstration area before beginning the demonstration. Do NOT substitute methyl alcohol or ethyl alcohol for isopropyl alcohol.

Perform the whoosh bottle demonstration in a plastic jug—never perform the reaction in a glass bottle. The large quantities of gases produced during the reaction may shatter a glass container. Inspect the plastic whoosh bottle before performing the demonstration—do not use the bottle if it shows any signs of frosting, cracking, or other flaws. Replace the whoosh bottle after 20 demonstrations.

Pour out any excess isopropyl alcohol from the whoosh bottle before igniting the gas mixture. Any liquid alcohol remaining in the jug will increase the amount of gaseous afterburning and could also ignite, causing the plastic bottle to melt. Excess alcohol remaining on the outside of the bottle should also be wiped off prior to igniting the alcohol vapor.

Be prepared:

- *Perform the demonstration behind a safety shield to prevent possible injury or damage from shattered objects.*

- *Have a fire blanket and fire extinguisher on hand to cover the bottle and extinguish the flame if the flame becomes too large or the reaction continues for too long.*

"Whoosh Bottle" is available as a Chemical Demonstration Kit from Flinn Scientific (Catalog No. AP5943).

- *The demonstrator and all spectators should be wearing protective eyewear.*

Please review current Material Safety Data Sheets for additional safety, handling, and disposal information for isopropyl alcohol.

Procedure

1. Measure 25 mL of isopropyl alcohol in a graduated cylinder and pour the alcohol into the special 5-gallon "whoosh bottle."

2. Recap the alcohol bottle and remove it from the demonstration area.

3. Lay the whoosh bottle sideways on a flat surface and allow the alcohol to flow from the base of the bottle to the mouth. Slowly swirl the bottle for about 30 seconds to spread the liquid completely over the entire interior surface. This allows the alcohol to evaporate more easily and makes the vapor concentration uniform throughout the bottle. If liquid is still visible, swirl the bottle for another 30 seconds.

4. Pour out any excess liquid from the whoosh bottle and shake out the bottle to remove the last traces of liquid.

5. Place the jug on the floor, in the front of the room and behind a safety shield.

6. Dim the lights in the room.

7. Tape a wood splint to a meter stick and light the splint with a match.

8. Stand back and, at arm's length, bring the burning splint over or slightly down into the mouth of the bottle.

9. Observe the explosive "whoosh" that results. The sound is produced by a moderately violent thrust of flames and blue gas out of the mouth of the bottle.

10. The whoosh is followed by a slower burn of gas down the inside surface of the bottle, producing a ring, plate, or cone of fire. This may be accompanied by an upward thrust or ball of yellow flames in the center of the jug.

11. Discuss the kinds of energy produced in this highly exothermic reaction: heat, light, and sound.

12. After the reaction has subsided and all the flames are out, wait for a minute or two to allow the bottle to cool slightly.

13. Feel the sides of the bottle to demonstrate the heat produced in an exothermic reaction.

14. Observe the whoosh bottle—it should contain a small amount of liquid. This is the water that is produced in the combustion reaction.

15. Using a small funnel, pour out the water from the whoosh bottle into a 25-mL graduated cylinder. Measure the volume of water produced—a typical yield is 12–14 mL of water.

16. Do NOT repeat the demonstration immediately! Please see the *Supplementary Information* section for advice on when and how the demonstration may be repeated.

Teacher Notes

Disposal

Excess isopropyl alcohol may be allowed to evaporate in a fume hood according to Flinn Suggested Disposal Method #18a. Consult your current *Flinn Scientific Catalog/Reference Manual* for general guidelines and specific procedures governing the disposal of laboratory waste.

Tips

- The best bottles for the "whoosh bottle" demonstration are the 5-gallon plastic polycarbonate containers commonly used to sell bottled water.

- The demonstration works best if the alcohol vapor is prepared immediately before the demonstration. If the bottle sits for a while, the vapor tends to settle and is harder to light.

- Depending on how much alcohol vapor is in the bottle, it may be necessary to place the flame slightly inside the lip of the whoosh bottle before it ignites.

- The whoosh bottle demonstration should only be performed with isopropyl alcohol. Ethyl alcohol is more volatile than isopropyl alcohol and thus reacts faster and more violently. n-Propyl alcohol burns more slowly and produces more heat, which may damage the bottle. Methyl alcohol is particularly dangerous—its high volatility may cause a violent explosion that will rupture the bottle.

- Use the water produced in the reaction to talk about the products of combustion. Discuss the formation of carbon dioxide as well and the danger of incomplete combustion, which can produce poisonous carbon monoxide. Use balanced chemical equations to demonstrate what kinds of conditions (insufficient oxygen) will lead to the formation of carbon monoxide. Relate these ideas to rules of fire safety.

Discussion

Low-boiling alcohols vaporize readily. When isopropyl alcohol is placed in a 5-gallon, small-mouthed jug, it forms a volatile mixture with the air. Only a small amount of alcohol is used and it quickly vaporizes to a heavier-than-air vapor. A simple flame held by the mouth of the jug provides the activation energy needed for the combustion of the alcohol/air mixture. Alcohol molecules in the vapor phase are farther apart than in the liquid phase and present a large surface area for reaction—the combustion reaction that occurs is therefore very fast. Since the burning is so rapid and occurs in the confined space of a 5-gallon jug with a small neck, the sound produced is very interesting, sounding like a "whoosh."

Equation 1 shows the balanced chemical equation for the combustion reaction of isopropyl alcohol:

$$(CH_3)_2CHOH(g) \ + \ 9/2O_2(g) \ \rightarrow \ 3CO_2(g) \ + \ 4H_2O(g) \ + \ heat \qquad \textit{Equation 1}$$

Use a graduated cylinder to measure the actual volume of water produced in the whoosh bottle reaction. Have students calculate the theoretical volume of water that should be produced from complete combustion of isopropyl alcohol, based on the amount of alcohol used in the demonstration. For example, if 20 mL of isopropyl alcohol (density = 0.78 g/mL) is used:

$$20 \text{ mL} \ \times \ \frac{0.78 \text{ g}}{1 \text{ mL}} \ \times \ \frac{1 \text{ mole}}{60 \text{ g}} \ = \ 0.26 \text{ mole isopropyl alcohol (IPA)}$$

$$0.26 \text{ mole IPA} \times \frac{4 \text{ mole H}_2\text{O}}{1 \text{ mole IPA}} \times \frac{18 \text{ g}}{1 \text{ mol}} = 18.7 \text{ g} = 18.7 \text{ mL of H}_2\text{O (theoretical yield)}$$

Compare the theoretical and actual yields of water. Discuss possible reasons why the actual yield of water is less than the maximum theoretical yield (evaporation due to the heat of reaction, the large surface area of the inside of the bottle, etc.).

Use standard heat of formation thermodynamic data for the products and reactants of the combustion reaction to calculate the theoretical heat of reaction (heat of combustion) for the complete combustion of isopropyl alcohol. The standard heat of reaction is calculated as the sum of the heats of formation of the products minus the sum of the heats of formation of the reactants, as shown in Equation 2. Notice that each term in the sum must be multiplied by its stoichiometric coefficient (n) from the balanced chemical equation.

$$\Delta H^\circ = n\Sigma\Delta H^\circ_f \text{ (products)} - n\Sigma\Delta H^\circ_f \text{ (reactants)} \qquad \textit{Equation 2}$$

Applying Equation 2 to the balanced chemical equation for the combustion of isopropyl alcohol (Equation 1) gives Equation 3:

$$\Delta H^\circ = 4\Delta H^\circ_f (\text{H}_2\text{O}) + 3\Delta H^\circ_f (\text{CO}_2) - 9/2\Delta H^\circ_f (\text{O}_2) - \Delta H^\circ_f (\text{C}_3\text{H}_7\text{OH}) \qquad \textit{Equation 3}$$

Since the heat of formation of any element in its standard state is zero, the oxygen term may be removed from the above equation, and Equation 3 reduces to Equation 4:

$$\Delta H^\circ = 4\Delta H^\circ_f (\text{H}_2\text{O}) + 3\Delta H^\circ_f (\text{CO}_2) - \Delta H^\circ_f (\text{C}_3\text{H}_7\text{OH}) \qquad \textit{Equation 4}$$

Substituting in the values for the heats of formation of water (–241.82 kJ/mole), carbon dioxide (–393.5 kJ/mole), and isopropyl alcohol (–272.2 kJ/mole) gives a calculated heat of combustion for the "whoosh bottle" reaction of –1,876 kJ/mol.

Supplementary Information

The whoosh bottle demonstration cannot and should not be repeated immediately. First of all, the demonstration will probably not work, due to the buildup of carbon dioxide in the bottle. There is not enough oxygen left in the bottle for combustion to occur. Secondly, it is dangerous to add additional isopropyl alcohol to the whoosh bottle if the bottle is still hot. In order to successfully repeat the demonstration, follow these steps:

1. Allow the whoosh bottle to cool to room temperature.

2. Pour out the water that forms as a result of the combustion.

3. Fill the bottle with about 1 inch of cold tap water and swirl the tap water around in the bottle. Pour the tap water into the sink, and repeat the process.

4. Dry out the bottle as much as possible by either allowing it to sit upside down or by drying it with a long string of paper towels pushed into the mouth.

5. In order to reduce the amount of water in the bottle and speed up the drying, try a double rinse of the bottle with a small amount of isopropyl alcohol.

Flameless Ration Heaters
An Applied Chemistry Demonstration

Introduction

Flameless ration heaters (FRH's) are nontoxic, self-contained heating units that do not produce a flame. FRH's were developed by the United States Army for use by soldiers in the field to heat their food rations. The use of flameless ration heaters in this demonstration provides an interesting, real-world application of thermochemistry.

Concepts

- Exothermic reaction
- Heat of reaction

Materials

Flameless ration heaters, 2	Erlenmeyer flask, 250 mL
Iron filings, Fe, 1 g	Filter funnel and filter paper
Magnesium ribbon, Mg, 2-cm strip	Forceps or tongs
Nitric acid, HNO_3, 3 M, 2 mL	Magnet, cow or rare-earth type
Phenolphthalein indicator solution, 2 mL	Magnifying glass
Silver nitrate solution, $AgNO_3$, 0.1 M, 1 mL	Safety shield
Sodium chloride, NaCl, 2 g	Scissors
Sodium hydroxide solution, NaOH, 0.1 M, 1 mL	Stirring rod
Balance	Test tubes, large (25 × 150 mm), 4
Bunsen burner	Thermometer
ChemCam™ Video Camera (optional)	Water
Deflagration spoon	

Safety Precautions

The Flameless Ration Heater is considered nontoxic and nonhazardous when used as described. The inner pad in the FRH consists of a supercorroding metal alloy embedded in a polyethylene matrix. The alloy consists of magnesium and about 5 mole percent iron. Handle the inner pad with forceps or tongs. Do not handle with bare hands. Magnesium metal is flammable and burns with an intense flame. Nitric acid is corrosive and a strong oxidizing agent; it is toxic by ingestion and inhalation. Sodium hydroxide solution is a skin and eye irritant. Silver nitrate solution will stain skin. Avoid contact of all chemicals with skin and eyes. Wear chemical splash goggles, and chemical-resistant gloves and apron. Consult current Material Safety Data Sheets for additional safety, handling, and disposal information.

Procedure

1. Obtain a FRH bag and display it to the class. Read the operating instructions and describe how the heater would be used in its intended application.

2. Cut the bag and remove the inner pad. Examine the pad with a magnifying lens and describe its appearance. *(Optional)* Place the pad under a ChemCam™ video camera to allow the class to see the magnified view.

Demonstrate the Reaction of the FRH Pad with Water:

3. Measure the mass of the inner metallic pad. Using scissors, cut a 3-cm² section of the FRH pad and record its mass. Fill a 250-mL Erlenmeyer flask with 100 mL of water and measure its initial temperature. Add the FRH section to the water and stir. Record the highest temperature that the water reaches.

4. *(Optional)* Calculate the amount of heat of reaction per gram of the pad. Multiply the heat of reaction per gram by the mass of the entire pad to calculate how much heat would be evolved when the entire pad is activated.

Demonstrate the Composition of the FRH Pad:

5. Suspend the inner pad vertically and bring a strong magnet near the pad. *(The particles in the pad are attracted to the magnet and will move toward the magnet. Carry out a parallel test with iron filings. Magnetism is a characteristic physical property of iron.)*

6. Filter the mixture from Step 3 through a funnel fitted with a filter paper cone. Collect two 10-mL fractions of the filtrate in large test tubes (25 × 150 mm).

7. Place a few drops of filtrate from the first test tube into a deflagration spoon and hold the spoon in a Bunsen burner flame. Describe the color of the flame. *(The flame should burn bright yellow-orange. Carry out a parallel flame test with sodium dissolved in water. A yellow-orange flame is characteristic of sodium ions.)*

8. To the remaining filtrate in the first test tube, carefully add 10 drops of 3 M nitric acid, followed by 3 drops of 0.1 M silver nitrate. Describe the observations. *(The filtrate is clear. Addition of silver nitrate gives a cloudy white mixture. Carry out a parallel test with a solution of sodium chloride dissolved in water. The formation of a white precipitate indicates the presence of chloride ions.)*

9. Add 2–3 drops of phenolphthalein indicator solution to the filtrate in the second test tube. Describe the observations. *(The filtrate turns red-violet. Carry out a parallel test with sodium hydroxide solution. The red-violet color indicates the presence of a strong base.)*

10. Cut a 1-cm² section of the FRH inner pad. Using tongs, hold the pad in a Bunsen burner flame behind a safety shield. The burning metal emits UV light that may harm the eyes—do NOT look directly at the flame. *(The pad burns with a brilliant light and an intense flame. Sparks fly off in several directions. Carry out a parallel test with a magnesium ribbon. The bright, intense flame is characteristic of magnesium metal.)*

Demonstrate the Reaction of Magnesium with Water:

11. Fill 2 large test tubes with about 10 mL of water. Add 2–3 drops of phenolphthalein to each.

12. Place a 2-cm strip of magnesium ribbon in each test tube. *(No reaction occurs.)*

13. To the second test tube, add a pea-sized amount of solid sodium chloride. Compare the reaction to that in the first test tube. *(Addition of sodium chloride increases the rate of reaction of magnesium with water. The mixture bubbles and turns pink, indicating the formation of hydrogen gas and basic magnesium hydroxide, respectively.)*

The Erlenmeyer flask is used in Step 3 to allow students to view the reaction as it occurs. To achieve more accurate measurements, however, the reaction should be carried out in an insulated foam cup.

Disposal

The waste solution remaining from reaction of the FRH pad with water is strongly basic and should be neutralized prior to disposal, according to Flinn Suggested Disposal Method #10. Please consult your current *Flinn Scientific Catalog/Reference Manual* for general guidelines and specific procedures governing the disposal of laboratory waste.

Tips

- Ask students to complete the following sentence describing the reaction of the FRH pad with water: "Reaction of the metallic pad with water is an *(exothermic/endothermic)* reaction. Heat is *(absorbed/released)* by the metallic pad and produces a large temperature *(increase/decrease)* in the surrounding water."

- Typical data for the amount of heat evolved in the reaction of the FRH pad with water: Reaction of 3.0 grams of the FRH pad with 100 mL of water increased the temperature of the water from 25 °C to 50 °C. This corresponds to about 3500 joules of heat per gram of FRH. The mass of a FRH pad is approximately 17 g, so the amount of heat that would be liberated when the entire pad is activated is about 60 kJ. The amount of heat released in this reaction is not very reproducible, probably because the pad itself is not homogeneous.

- Lead students on a chemical detective mission as you demonstrate the composition and reaction of the FRH pad. After each step, ask students leading questions to deduce what materials are present and how they react. What two metals are present in the pad? How do you know? What is the role of sodium chloride? What is the nature of the products that are formed?

Discussion

The metallic pad in the Flameless Ration Heater is a composite material containing a magnesium–iron alloy and sodium chloride dispersed in a polyethylene matrix. According to the patent for its manufacture, the FRH is capable of increasing the temperature of 151 grams of food approximately 55 °C—from an initial temperature of 21 °C, for example, to a final temperature of 76 °C. Some heat escapes the package in the form of steam and hydrogen gas, but the hydrogen gas dissipates rapidly and does not present a hazard. According to the patent, the hydrogen gas does not ignite, either with an open flame or with a spark.

Reaction of the FRH pad with water involves oxidation of magnesium to form magnesium hydroxide and hydrogen, according to the following balanced chemical equation:

$$Mg(s) + 2H_2O(l) \rightarrow Mg(OH)_2(s) + H_2(g) + heat \qquad Equation~1$$

The heat of reaction for this highly exothermic reaction is –352 kJ/mole of Mg. Although thermodynamically favored, the reaction of magnesium with water is kinetically very slow in the absence of a promoter and/or catalyst. Elemental iron is added to the metal as a promoter—it initiates reaction at the magnesium surface by producing reactive intermediates. The reaction further requires a catalyst to increase the rate of the reaction and ensure a smooth, steady evolution of heat. Sodium chloride is added to the metal composite material to break down or destabilize the protective coating on the surface of the magnesium metal. Chloride ions replace hydroxide ions in the coating and produce channels that allow water to penetrate the reactive metal surface. Chloride ions redissolve in solution and thus function as a true catalyst.

Safety and Disposal Guidelines

Safety Guidelines

Teachers owe their students a duty of care to protect them from harm and to take reasonable precautions to prevent accidents from occurring. A teacher's duty of care includes the following:

- Supervising students in the classroom.

- Providing adequate instructions for students to perform the tasks required of them.

- Warning students of the possible dangers involved in performing the activity.

- Providing safe facilities and equipment for the performance of the activity.

- Maintaining laboratory equipment in proper working order.

Safety Contract

The first step in creating a safe laboratory environment is to develop a safety contract that describes the rules of the laboratory for your students. Before a student ever sets foot in a laboratory, the safety contract should be reviewed and then signed by the student and a parent or guardian. Please contact Flinn Scientific at 800-452-1261 or visit the Flinn Website at www.flinnsci.com to request a free copy of the Flinn Scientific Safety Contract.

To fulfill your duty of care, observe the following guidelines:

1. **Be prepared.** Practice all experiments and demonstrations beforehand. Never perform a lab activity if you have not tested it, if you do not understand it, or if you do not have the resources to perform it safely.

2. **Set a good example.** The teacher is the most visible and important role model. Wear your safety goggles whenever you are working in the lab, even (or especially) when class is not in session. Students learn from your good example—whether you are preparing reagents, testing a procedure, or performing a demonstration.

3. **Maintain a safe lab environment.** Provide high-quality goggles that offer adequate protection and are comfortable to wear. Make sure there is proper safety equipment in the laboratory and that it is maintained in good working order. Inspect all safety equipment on a regular basis to ensure its readiness.

4. **Start with safety.** Incorporate safety into each laboratory exercise. Begin each lab period with a discussion of the properties of the chemicals or procedures used in the experiment and any special precautions—including goggle use—that must be observed. Pre-lab assignments are an ideal mechanism to ensure that students are prepared for lab and understand the safety precautions. Record all safety instruction in your lesson plan.

5. **Proper instruction.** Demonstrate new or unusual laboratory procedures before every activity. Instruct students on the safe way to handle chemicals, glassware, and equipment.

6. **Supervision.** Never leave students unattended—always provide adequate supervision. Work with school administrators to make sure that class size does not exceed the capacity of the room or your ability to maintain a safe lab environment. Be prepared and alert to what students are doing so that you can prevent accidents before they happen.

7. **Understand your resources.** Know yourself, your students, and your resources. Use discretion in choosing experiments and demonstrations that match your background and fit within the knowledge and skill level of your students and the resources of your classroom. You are the best judge of what will work or not. Do not perform any activities that you feel are unsafe, that you are uncomfortable performing, or that you do not have the proper equipment for.

Safety Precautions

Specific safety precautions have been written for every experiment and demonstration in this book. The safety information describes the hazardous nature of each chemical and the specific precautions that must be followed to avoid exposure or accidents. The safety section also alerts you to potential dangers in the procedure or techniques. Regardless of what lab program you use, it is important to maintain a library of current Material Safety Data Sheets for all chemicals in your inventory. Please consult current MSDS for additional safety, handling, and disposal information.

Disposal Procedures

The disposal procedures included in this book are based on the Suggested Laboratory Chemical Disposal Procedures found in the *Flinn Scientific Catalog/Reference Manual*. The disposal procedures are only suggestions—do not use these procedures without first consulting with your local government regulatory officials.

Many of the experiments and demonstrations produce small volumes of aqueous solutions that can be flushed down the drain with excess water. Do not use this procedure if your drains empty into groundwater through a septic system or into a storm sewer. Local regulations may be more strict on drain disposal than the practices suggested in this book and in the *Flinn Scientific Catalog/Reference Manual*. You must determine what types of disposal procedures are permitted in your area—contact your local authorities.

Any suggested disposal method that includes "discard in the trash" requires your active attention and involvement. Make sure that the material is no longer reactive, is placed in a suitable container (plastic bag or bottle), and is in accordance with local landfill regulations. Please do not inadvertently perform any extra "demonstrations" due to unpredictable chemical reactions occurring in your trash can. Think before you throw!

Finally, please read all the narratives before you attempt any Suggested Laboratory Chemical Disposal Procedure found in your current *Flinn Scientific Catalog/Reference Manual*.

Flinn Scientific is your most trusted and reliable source of reference, safety and disposal information for all chemicals used in the high school science lab. To request a complementary copy of the most recent *Flinn Scientific Catalog/Reference Manual,* call us at 800-452-1261 or visit our website at www.flinnsci.com.

Experiments and Demonstrations

Content Standards	Exploring Energy	Measuring Energy	Discovering Instant Cold Packs	Measuring Calories	Heat of Reaction and Hess's Law	Colorful Heat	Specific Heat	Cool Reaction	Whoosh Bottle	Flameless Ration Heaters
Unifying Concepts and Processes										
Systems, order, and organization	✓	✓	✓	✓	✓		✓	✓		
Evidence, models, and explanation	✓	✓	✓	✓	✓	✓	✓			✓
Constancy, change, and measurement	✓	✓	✓	✓	✓		✓	✓	✓	✓
Evolution and equilibrium		✓		✓	✓	✓	✓			
Form and function			✓							✓
Science as Inquiry										
Identify questions and concepts that guide scientific investigation			✓			✓	✓			✓
Design and conduct scientific investigations	✓	✓	✓	✓	✓		✓			✓
Use technology and mathematics to improve scientific investigations	✓	✓	✓	✓	✓		✓			✓
Formulate and revise scientific explanations and models using logic and evidence							✓			✓
Recognize and analyze alternative explanations and models					✓					
Communicate and defend a scientific argument			✓							
Understanding scientific inquiry	✓	✓	✓	✓	✓	✓	✓	✓	✓	✓
Physical Science										
Structure of atoms										
Structure and properties of matter	✓	✓	✓				✓			
Chemical reactions	✓			✓	✓			✓	✓	✓
Motions and forces										
Conservation of energy and the increase in disorder		✓	✓	✓	✓		✓	✓		✓
Interactions of energy and matter	✓	✓	✓	✓	✓		✓	✓	✓	✓

Content Standards *(continued)*	Exploring Energy	Measuring Energy	Discovering Instant Cold Packs	Measuring Calories	Heat of Reaction and Hess's Law	Colorful Heat	Specific Heat	Cool Reaction	Whoosh Bottle	Flameless Ration Heaters
Science and Technology										
Identify a problem or design an opportunity			✓							
Propose designs and choose between alternative solutions			✓	✓	✓		✓			
Implement a proposed solution			✓							
Evaluate the solution and its consequences			✓							
Communicate the problem, process, and solution										
Understand science and technology			✓	✓						✓
Science in Personal and Social Perspectives										
Personal and community health			✓							✓
Population growth										
Natural resources										
Environmental quality										
Natural and human-induced hazards									✓	
Science and technology in local, national, and global challenges				✓						✓
History and Nature of Science										
Science as a human endeavor			✓	✓						✓
Nature of scientific knowledge	✓	✓	✓	✓	✓	✓	✓	✓	✓	✓
Historical perspectives										

Experiments and Demonstrations

(for a class of 30 students working in pairs)   **Experiments and Demonstrations***

	Flinn Scientific Catalog No.	Exploring Energy	Measuring Energy	Discovering Instant Cold Packs	Measuring Calories	Heat of Reaction and Hess's Law	Colorful Heat	Specific Heat	Cool Reaction	Whoosh Bottle	Flameless Ration Heaters
Chemicals											
Aluminum shot	A0262							60 g			
Ammonium chloride	A0045	80 g									
Ammonium nitrate	A0056			225 g							
Ammonium thiocyanate	A0213								10 g		
Barium hydroxide octahydrate	B0155								20 g		
Calcium chloride	C0016	110 g									
Hydrochloric acid solution, 1 M	H0057	300 mL				900 mL					
Iron filings	I0011										1 g
Isopropyl alcohol	I0019									25 mL	
Lead shot	L0090							60 g			
Magnesium oxide	M0013					6 g					
Magnesium ribbon	M0139					105 cm					2 cm
Nitric acid, 3 M	N0049										2 mL
Phenolphthalein solution, 1%	P0019										2 mL
Silver nitrate solution, 0.1 M	S0305										1 mL
Sodium bicarbonate	S0043	40 g									
Sodium chloride	S0063										2 g
Sodium hydroxide solution, 0.1 M	S0149										1 mL
Zinc shot	Z0021							60 g			
Glassware											
Beakers											
250-mL	GP1020							5			
400-mL	GP1025	15	30	15					3		
600-mL	GP1030								1		
Erlenmeyer flasks											
125-mL	GP3040				15				1		
250-mL	GP3045										1

*Many of these experiments and demonstrations will be adapted into student laboratory kits.
Consult your current *Flinn Scientific Catalog/Reference Manual* for kit availability.

(for a class of 30 students working in pairs)   **Experiments and Demonstrations**

	Flinn Scientific Catalog No.	Exploring Energy	Measuring Energy	Discovering Instant Cold Packs	Measuring Calories	Heat of Reaction and Hess's Law	Colorful Heat	Specific Heat	Cool Reaction	Whoosh Bottle	Flameless Ration Heaters
Glassware, continued											
Graduated cylinders											
10-mL	GP2005	15									
25-mL	GP2010					15				1	
50-mL	GP2015	15			15						
100-mL	GP2020		15	15				1			
250-mL	GP2025		15								
Test tubes											
25 × 150 mm	GP6035							3			4
Stirring rods	GP5075	15	15	15	15	15	1	3	1		1
General Equipment and Miscellaneous											
Bags, zipper-lock, plastic	AB1004	15									
Balance, centigram (0.01-g precision)	OB2059	3		3	3	3		1	1		1
Boiling stones	B0136							3			
Bunsen burner	AP8344										1
Calorimeter, metal, economy choice*	AP4533				15						
Calorimeter, small-scale	AP5928					15					
ChemCam video camera	AP4560									optional	
Cold pack, instant	AP6267		2								
Deflagration spoon	AP1346										1
Filter paper	AP3104										1
Fire blanket	SE3006									1	
Fire extinguisher	SE3001									1	
Flameless ration heater	AP8695										2
Food coloring	V0003						1				
Forceps	AP8328		15			15					1
Funnel	AP3200									1	1
Gloves, Zetex,™ for low temperature	AP3240								1		

* AP4533 includes the calorimeter and lid, Erlenmeyer flask with collar, and food holder and pin.

Continued on next page

(for a class of 30 students working in pairs) **Experiments and Demonstrations**

General Equipment and Miscellaneous, cont'd.	Flinn Scientific Catalog No.	Exploring Energy	Measuring Energy	Discovering Instant Cold Packs	Measuring Calories	Heat of Reaction and Hess's Law	Colorful Heat	Specific Heat	Cool Reaction	Whoosh Bottle	Flameless Ration Heaters
Hot Hands Heat Protector	SE039		15								
Hot plate	AP4674		5–8					1			
LabPro Interface system	TC1500	optional	optional	optional	optional	optional					
LoggerPro Software	TC1421	optional	optional	optional	optional	optional					
Magnet	AP9041										1
Magnifier	AP6020										1
Marking pencil or pen	AP8921	15						1			
Matches	AP1935				15					1	
Meter stick	AP5384									1	
Metric rulers	AP4684					15					
Rubber stopper	AP2320								1		
Safety shield	SE225									1	1
Scissors	AP4396					3					1
Spatula	AP8336	15		15		15					
Specific Heat Set	AP9220							1			
Styrofoam® cups, 6 oz	AP1190	15	15	30				6			
Test tube clamps	AP8217							3			
Thermometer, digital —or— Temperature probe	AP4852 —or— TC1502	15	15	15	15	15		3	1		1
Wash bottle	AP1668					15			1		
Water, distilled or deionized	W0007	✓	✓	✓		✓		✓		✓	✓
Weighing dishes	AP1278	60		15		30					
Whoosh Bottle Kit	AP5943									1	
Wood splints	AP4444									1	